The ESS

PHYSICAL CHEMISTRY II

Staff of Research and Education Association,
Dr. M. Fogiel, Director

This book is a continuation of *"THE ESSENTIALS OF PHYSICAL CHEMISTRY I"* and begins with Chapter 15. It covers the usual course outline of Physical Chemistry II. Earlier/basic topics are covered in *"THE ESSENTIALS OF PHYSICAL CHEMISTRY I"*.

Research and Education Association
61 Ethel Road West
Piscataway, New Jersey 08854

THE ESSENTIALS OF PHYSICAL CHEMISTRY II

Printed in the United States of America

Library of Congress Catalog Card Number 87-61801

International Standard Book Number 0-87891-621-0

Revised Printing 1989

WHAT "THE ESSENTIALS" WILL DO FOR YOU

This book is a review and study guide. It is comprehensive and it is concise.

It helps in preparing for exams, in doing homework, and remains a handy reference source at all times.

It condenses the vast amount of detail characteristic of the subject matter and summarizes the **essentials** of the field.

It will thus save hours of study and preparation time.

The book provides quick access to the important facts, principles, theorems, concepts, and equations of the field.

Materials needed for exams, can be reviewed in summary form – eliminating the need to read and re-read many pages of textbook and class notes. The summaries will even tend to bring detail to mind that had been previously read or noted.

This "ESSENTIALS" book has been carefully prepared by educators and professionals and was subsequently reviewed by another group of editors to assure accuracy and maximum usefulness.

Dr. Max Fogiel
Program Director

CONTENTS

This book is a continuation of *"THE ESSENTIALS OF PHYSICAL CHEMISTRY I"* and begins with Chapter 15. It covers the usual course outline of Physical Chemistry II. Earlier/basic topics are covered in *"THE ESSENTIALS OF PHYSICAL CHEMISTRY I"*.

Chapter No. **Page No.**

15 REACTION MECHANISMS 118

15.1 Mechanism of a Reaction 118
15.2 Opposing Reactions and Equilibrium Constants 119
15.3 Consecutive and Parallel Reactions and Steady State 121
15.4 Complex Reactions 122

16 THEORETICAL APPROACHES TO CHEMICAL KINETICS 124

16.1 Temperature Dependence of Reaction Rate 124
16.2 The Collision Theory 125
16.3 The Activated Complex Theory 127
16.4 Unimolecular Reaction and the Lindemann Theory 129

17	**GRAVITATIONAL, ELECTRICAL AND MAGNETIC WORK**	131
17.1	Work and Energy	131
17.2	Gravitational Work	133
17.3	Electrical Work and the Piezoelectric Effect	134
17.4	Magnetic Work and Adiabatic Cooling	135
18	**SURFACE WORK**	140
18.1	Surface Tension and Surface Energy	140
18.2	Bubbles and Drops	141
18.3	The Kelvin Equation	143
18.4	Gibbs Formulation of Adsorption	145
18.5	The Langmuir Adsorption Isotherm	146
18.6	Statistical Mechanics of Adsorption	147
18.7	The Ising Adsorption Model	150
19	**KINETIC THEORY**	153
19.1	Statistical Methods	153
19.2	The Binomial Distribution	153
19.3	The Gaussian Distribution	154
19.4	The Boltzmann Distribution	156
19.5	One-Dimensional Velocity Distribution	156
19.6	Some Useful Integrals	159
19.7	Higher Dimensional Distributions	159
19.8	The Average Kinetic Energy and the Most Probable Speed	162
19.9	The Pressure of a Gas	163
20	**COLLISIONAL AND TRANSPORT PROPERTIES OF GASES**	165
20.1	Approximate Solution of Molecular Effusion	165

20.2 Exact Solution of Molecular Effusion 167
20.3 Mean Free Path and Collision Frequency 169
20.4 Viscosity 174
20.5 Thermal Conductivity 177
20.6 Diffusion 177

21 STATISTICAL MECHANICS 180

21.1 Entropy and Disorder 180
21.2 Lagrange's Method for Constrained Extrema and the Stirling Approximation 181
21.3 Ensembles 182
21.4 The Boltzmann Distribution 184
21.5 Equations of State and the Molecular Partition Function 189
21.6 Evaluating the Molar Partition Function 190
21.7 The Translational Partition Function 192

22 MATTER AND WAVES 195

22.1 Simple Harmonic Motion 195
22.2 Wave Motion 197
22.3 Standing Waves 198
22.4 Blackbody Radiation and the Nuclear Atom 200
22.5 The Photoelectric Effect 203
22.6 Spectroscopy and the Bohr Atom 203
22.7 The de Broglie Relation 206

23 QUANTUM MECHANICS 209

23.1 The Schrödinger Equation 209
23.2 Postulates of Quantum Mechanics 211
23.3 Operators 213
23.4 Solutions of Schödinger's Equation 214
23.4.1 The Free Particle 214

23.4.2 The Particle in a Ring of Constant Potential 215
23.4.3 The Particle in a Box 217
23.4.4 The One-Dimensional Box with One Finite Wall 222

24 ROTATIONS AND VIBRATIONS OF ATOMS AND MOLECULES 226

24.1 Harmonic Oscillator 226
24.2 The Nature of the Harmonic Oscillator Wavefunctions 230
24.3 Thermodynamics of the Harmonic Oscillator Wavefunctions 231
24.4 The Rigid Diatomic Rotor 233
24.5 The Thermodynamics of the Rigid Rotor 236

PERIODIC TABLE 238

CHAPTER 15

REACTION MECHANISMS

15.1 MECHANISM OF A REACTION

The mechanism of a reaction is the path that the reaction takes. The following sequence of reactions is the mechanism of the reaction

$$A_2B_5 \rightarrow A_2B_4 + \tfrac{1}{2}B_2$$

$$A_2B_5 \rightarrow AB_2 + AB_3 \qquad (a)$$

$$AB_2 + AB_3 \rightarrow A_2B_5 \qquad (b)$$

$$AB_2 + AB_2 \rightarrow AB_2 + \tfrac{1}{2}B_2 + AB \qquad (c)$$

$$AB + A_2B_5 \rightarrow 3AB_2 \qquad (d)$$

$$AB_2 + AB_2 \rightarrow A_2B_4 \qquad (e)$$

Each of the above intermediate reactions is an elementary reaction. Molecular events are described by the elementary reactions.

The rate equation for a complex mechanism is the sum of the rate equations for the simple reactions composing it.

$$A + B \underset{k_{-1}}{\overset{k_1}{\rightleftharpoons}} C \qquad\qquad C + B \xrightarrow{k_2} D$$

$$\frac{-dC_A}{dt} = k_1 C_A C_B - k_{-1} C_C$$

$$\frac{-dC_B}{dt} = k_1 C_A C_B - k_{-1} C_C + k_2 C_C C_B$$

$$\frac{+dC_C}{dt} = k_1 C_A C_B - k_{-1} C_C - k_2 C_C C_B$$

15.2 OPPOSING REACTIONS AND EQUILIBRIUM CONSTANTS

The rate equation for the first order reaction

$$A \underset{k_{-1}}{\overset{k_1}{\rightleftharpoons}} B$$

is

$$\frac{-dC_A}{dt} = \frac{dC_B}{dt} = k_1 C_A - k_{-1} C_B$$

where k_1 = the forward rate constant

k_{-1} = the reverse rate constant.

At equilibrium, $\frac{-dC_A}{dt} = 0 = \frac{dC_B}{dt}$

and

$$\frac{k_1}{k_{-1}} = \frac{C_B}{C_A} = K$$

where K is the equilibrium constant and is known as the principle of Guldberg and Waage, or the law of mass action.

$$\frac{dx}{dt} = k_1(C_{A_0} - x) - k_{-1}(C_{B_0} + x)$$

where C_{A_0} = the initial concentration of A

C_{B_0} = the initial concentration of B

x = the amount of A that has been transformed into B.

$$\frac{dx}{dt} = (k_1 + k_{-1})(C - x)$$

where $$C = \left(\frac{k_1 C_{A_0} - k_{-1} C_{B_0})}{k_1 + k_{-1}} \right)$$

$$\ln\left(\frac{C}{C - X}\right) = (k_1 + k_{-1})t$$

For the reaction $A + B \underset{k_{-2}}{\overset{k_2}{\rightleftharpoons}} 2C$, the rate equation is

$$\frac{dC_C}{dt} = -2\frac{dC_A}{dt} = -2\frac{dC_B}{dt} = \frac{dx}{dt}$$

$$\frac{dx}{dt} = k_2(C_{A_0} - \tfrac{1}{2}x)(C_{B_0} - \tfrac{1}{2}x) - k_{-2}x^2$$

where C_{A_0} = the initial concentration of A
C_{B_0} = the initial concentration of B
x = the amount of C formed.

$$\ln\left\{\left[\frac{(C_{A_0}+C_{B_0}+D)/(1-4/K)-X}{(C_{A_0}+C_{B_0}+D)/(1-4/K)-X}\right]\left[\frac{C_{A_0}+C_{B_0}+D}{C_{A_0}+C_{B_0}+D}\right]\right\} = \tfrac{1}{2}(1-4K)Dk_2t$$

where $D = \sqrt{(C_{A_0}+C_{B_0})^2-4C_{A_0}C_{B_0}(1-4K)^{-1}}$ and

$$K = \frac{k_2}{k_{-2}} = \text{equilibrium constant.}$$

$$K = \frac{k_2}{k_{-2}} = \frac{C_C^2}{C_A C_B}$$

The principle of detailed balancing assumes that the rate of an elementary process is equal to the rate of the reverse process at equilibrium. The principle of detailed balancing is an extension of the principle of microscopic reversibility, which states that all the momenta (both translational and internal) are reversed when two molecules collide and then the system returns by the same process in the reverse direction.

15.3 CONSECUTIVE AND PARALLEL REACTIONS AND STEADY STATE

CONSECUTIVE REACTIONS

For the consecutive first-order reactions,

$$A \xrightarrow{k_1} B \qquad B \xrightarrow{k_2} C$$

the concentrations at time t are

$$C_A = C_{A,O}\, e^{-k_1 t}$$

$$C_B = \frac{k_1 C_{A,O}}{k_2 - k_1}\left(e^{-k_1 t} - e^{-k_2 t}\right)$$

$$C_C = C_{A,O}\left(1 - \frac{k_2 e^{-k_1 t} - k_1 e^{-k_2 t}}{k_2 - k_1}\right)$$

$$C_{A,O} = C_A + C_B + C_C \qquad \text{when } C_{B,O} = C_{C,O} = 0$$

For the two consecutive reversible reactions,

$$A \underset{k_{-1}}{\overset{k_1}{\rightleftharpoons}} B \underset{k_{-2}}{\overset{k_2}{\rightleftharpoons}} C$$

at equilibrium

$$\left(\frac{C_B}{C_A}\right)_{eq} = \frac{k_1}{k_{-1}} \qquad \left(\frac{C_C}{C_B}\right)_{eq} = \frac{k_2}{k_{-2}} \qquad \left(\frac{C_C}{C_A}\right)_{eq} = \frac{k_1 k_2}{k_{-1}k_{-2}}$$

Parallel reactions

For the parallel reactions

$$A + B \xrightarrow{k_1} C + D \qquad A + B \xrightarrow{k_2} 2E \qquad A + B \xrightarrow{k_3} F + G$$

$$\text{rate} = k_1 C_A C_B + k_2 C_A C_B + k_3 C_A C_B = k C_A C_B$$

where $\qquad k = k_1 + k_2 + k_3$

STEADY STATE

The steady state approximation is the assumption that the reactive intermediate substances are at constant concentrations as the reaction takes place.

For the reactions

$$A + B \underset{k_{-1}}{\overset{k_1}{\rightleftharpoons}} C \qquad C + B \xrightarrow{k_2} D$$

$$\frac{dC_C}{dt} = k_1 C_A C_B - k_{-1} C_C - k_2 C_C C_B$$

$$C_C = \frac{k_1 C_A C_B}{k_{-1} + k_2 C_B} \qquad \text{when} \quad \frac{dC_C}{dt} = 0$$

$$\frac{-dC_A}{dt} = k_1 C_A C_B - \frac{k_{-1} k_1 C_A C_B}{k_{-1} + k_2 C_B} = \frac{k_1 k_2 C_A C_B^2}{k_{-1} + k_2 C_B}$$

$$\frac{-dC_B}{dt} = k_1 C_A C_B - \frac{k_{-1} k_1 C_A C_B}{k_{-1} + k_2 C_B} + \frac{k_2 C_B k_1 C_A C_B}{k_{-1} + k_2 C_B}$$

$$\frac{-dC_B}{dt} = \frac{2 k_1 k_2 C_A C_B^2}{k_{-1} + k_2 C_B}$$

$$\frac{dC_D}{dt} = \frac{k_1 k_2 C_A C_B^2}{k_{-1} + k_2 C_B}$$

15.4 COMPLEX REACTIONS

A chain reaction is the sequence of reactions, in the gas phase, of radicals producing radicals.

The chain reaction is analyzed by the following steps:

1) Initiation step, in which normal molecules form radicals.

Example:

$$Br_2 \xrightarrow{k_1} 2Br\cdot$$

2) Propagation step, in which the free radicals attack other molecules to produce new radicals. Example:

$$Br\cdot + H_2 \xrightarrow{k_2} HBr + H\cdot$$

$$H\cdot + Br_2 \xrightarrow{k_3} HBr + Br\cdot$$

3) Inhibition step, in which the radical attacks the product molecule. Example:

$$H\cdot + HBr \xrightarrow{k_4} H_2 + Br\cdot$$

4) Termination step, in which the chain of radical reactions ends. Example:

$$Br\cdot + Br\cdot \xrightarrow{k_5} Br_2$$

$$\frac{d[HBr]}{dt} = k_2[Br][H_2] + k_3[H][Br_2] - k_4[H][HBr]$$

$$\frac{d[H]}{dt} = k_2[Br][H_2] - k_3[H][Br_2] - k_4[H][HBr] \sim 0$$

$$\frac{d[Br]}{dt} = 2k_1[Br_2] - k_2[Br][H_2] + k_3[H][Br_2]$$

$$+ k_4[H][HBr] - 2k_5[Br]^2 \sim 0$$

$$\frac{d[HBr]}{dt} = \frac{2k_2(k_1/k_5)^{\frac{1}{2}}[H_2][Br_2]^{\frac{1}{2}}}{1 + (k_4[HBr]/k_3[Br_2])}$$

CHAPTER 16

THEORETICAL APPROACHES TO CHEMICAL KINETICS

16.1 TEMPERATURE DEPENDENCE OF REACTION RATE

The reaction rate increases as the temperature increases. For every 10°K increase in temperature, the reaction rate doubles. The rate constant is related to the temperature by the following Arrhenius equation:

$$\boxed{k = A \exp\left(\frac{-E_a}{RT}\right)}$$

$$\ln k = \ln A \frac{-E_a}{RT}$$

where
k = the rate constant

A = the frequency factor or the pre-exponential factor

E_a = the activation energy

T = the temperature in °K.

$$\ln \frac{k_2}{k_1} = -\left(\frac{\Delta E_a}{R}\right)\left(\frac{1}{T_2} - \frac{1}{T_1}\right)$$

$$\frac{d(\ln K)}{dt} = \frac{\Delta E}{RT^2}$$

where K is the equilibrium constant and ΔE is the change in energy.

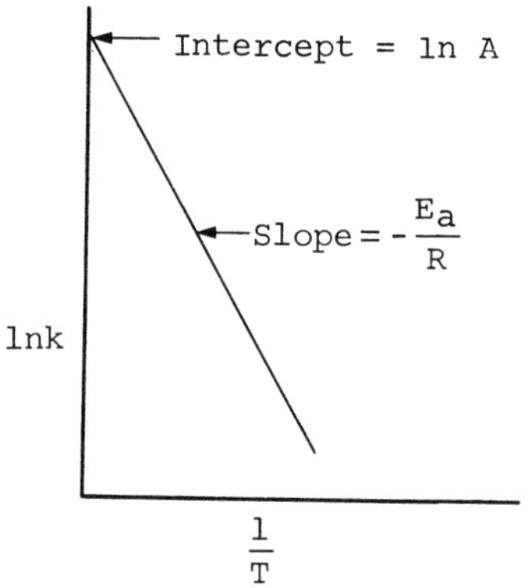

Fig. 16.1 Arrhenius plot that shows the relationship between the rate constant and the temperature.

16.2 THE COLLISION THEORY

The collision theory postulates that in order for molecules to interact, they must approach each other so closely that they can be said to be in collision. The speed of reaction is equal to the number of collisions per second times the fraction of the collisions that are effective in producing chemical change.

The collision between like molecules is

$$Z_{AA} = \tfrac{1}{2}\sqrt{2}\ \pi d^2 \bar{c}(n^*)^2$$

where $\bar{c} = \sqrt{\dfrac{8kT}{\pi m}}$ = the average velocity

n^* = the number of gas molecules per cubic centimeter

m = the mass

d = the diameter of the molecules

Z_{AA} = the collision frequency.

The collision between unlike molecules is

$$Z_{AB} = n_A{}^* n_B{}^* \; d^2AB \left[8\pi RT \frac{m_A + m_B}{m_A m_B} \right]^{\frac{1}{2}}$$

where $n_A{}^*$ and $n_B{}^*$ = the concentrations of A and B

d_{AB} = the average of the two diameters (or the sum of the molecular radii)

$\left(\frac{m_A + m_B}{m_A m_B} \right)$ = the reduced mass of the two molecules.

d^2_{AB} = the collision cross section which corresponds to the area of the molecules for a collision.

The rate of reaction for unlike colliding molesules is

$$\text{rate} = 4P \left(\frac{\pi kT}{m} \right)^{\frac{1}{2}} d^2(10^3L)e^{-\Delta E_a/RT} \; C^2$$

The rate of reaction for like colliding molecules is

$$\text{rate} = P \left[8\pi kT \frac{m_A + m_B}{m_A m_B} \right]^{\frac{1}{2}} d^2_{AB}(10^3L)e^{-\Delta E_a/RT} \; C_A C_B$$

where P is the steric factor which depends upon the relative position of the colliding molecules.

The Arrhenius pre-exponential factor for like molecules is

$$A = 4P \left(\frac{\pi kT}{m} \right)^{\frac{1}{2}} d^2(10^3L)$$

and for unlike molecules is

$$A = P \left[8\pi kT \frac{m_A + m_B}{m_A m_B} \right]^{\frac{1}{2}} d^2_{AB}(10^3 L)$$

16.3 THE ACTIVATED COMPLEX THEORY

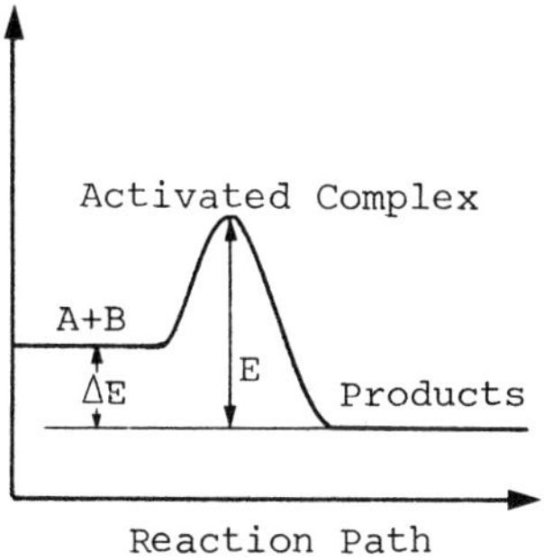

Fig. 16.2 The change of energy throughout a reaction

The activated complex passes through a transition state with a maximum energy. The activated complex theory is called the theory of absolute reaction rates.

$$A + B \underset{}{\overset{K^{\ddagger}}{\rightleftharpoons}} C^{\ddagger} \xrightarrow{k} \text{products}$$

$$K^{\ddagger} = \frac{[C^{\ddagger}]}{[A][B]} = \frac{Q^{\ddagger}}{Q_A Q_B} \exp\left(\frac{-E}{RT}\right)$$

where

$K^{\ddagger}$ = the equilibrium constant for the activated complex

$Q^{\ddagger}$ = the partition function for the activated complex

Q_A and Q_B = the partition functions for the reactants A and B.

E = the value shown in Figure 16.2.

$$\text{rate} = \kappa \left(\frac{kT}{h}\right)\left(\frac{Q^{\ddagger}}{Q_A Q_B}\right) \exp\left(\frac{-E}{RT}\right) C_A C_B$$

where κ (kappa) = the transmission coefficient

$$Q_i = q_{trans} q_{rot} q_{vib}$$

where q_{trans}, q_{rot} and q_{vib} are the partition functions.

$$q_{trans} = (2\pi mkT)^{\frac{3}{2}} \frac{V}{h^3}$$

$$q_{rot} = \frac{8\pi^2 IkT}{h^2 \sigma}$$

$$q_{vib} = \prod_{i=1}^{3\Lambda-5 \text{ or } 3\Lambda-6} \frac{1}{1 - e^{-x_i}}$$

$3\Lambda - 5$ is the degrees of freedom for a linear reactant, and $3\Lambda - 6$ for a non-linear reactant.

$$Q^{\ddagger} = (2\pi m^{\ddagger} kT)^{\frac{5}{2}} \frac{4\pi m_A m_B \sigma_{AB}^2}{m^{\ddagger 2} h^5}$$

$$I = \frac{m_A m_B \sigma_{AB}^2}{m^{\ddagger}}, \quad m^{\ddagger} = m_A + m_B \text{ and } Q^{\ddagger} = q_{trans} q_{rot}$$

$$Q_A = \frac{(2\pi m_A kT)^{\frac{3}{2}}}{h^3} \qquad Q_B = \frac{(2\pi m_B kT)^{\frac{3}{2}}}{h^3}$$

where I corresponds to the moment of inertia and σ_{AB} to the bond length of the activated complex.

The activated complex theory and thermodynamic considerations.

$$\Delta G^{\circ\ddagger} = \Delta H^{\circ\ddagger} + T\Delta S^{\circ\ddagger} = -RT \ln K^{\ddagger}$$

$$\text{rate} = \kappa \frac{kT}{h} e^{\Delta S^{\circ\ddagger}/R} e^{-\Delta H^{\circ\ddagger}/RT} C_A C_B$$

where $\Delta G^{\circ\ddagger}$ = the Gibbs energy of activation

$\Delta S^{\circ\ddagger}$ = the entropy of activation

$\Delta H^{\circ\ddagger}$ = the enthalpy of activation.

$$A = e^n \left(\frac{kT}{h}\right) e^{\Delta S^{\circ\ddagger}/R}$$

where A is the pre-exponential factor.

$$\Delta H^{\circ\ddagger} = \Delta E_a - nRT$$

where ΔE_a is the activation energy.

16.4 UNIMOLECULAR REACTION AND THE LINDEMANN THEORY

The Lindemann theory is used to explain the first-order reactions by the following two-step mechanism:

1) $A + A \underset{k_{-2}}{\overset{k_2}{\rightleftarrows}} A^* + A$

2) $A^* \xrightarrow{k_1}$ products (P)

A is a normal reactant molecule and A* is an energized or activated molecule.

By applying the steady-state approximation

$$[A^*] = \frac{k_2[A]^2}{k_{-2}[A] + k_1}$$

$$\frac{d[P]}{dt} = k_1[A^*] = \frac{k_1 k_2[A]^2}{k_{-2}[A] + k_1}$$

$$\frac{d[P]}{dt} = \frac{k_1 k_2 [A]}{k_{-2}} = k_\infty [A] \text{ at high pressure}$$

$$\frac{d[P]}{dt} = k_2[A]^2 \text{ at low pressure.}$$

The system changes gradually from first-order to second-order as the pressure is lowered.

$$\frac{1}{k_{exp}} = \frac{1}{k_{\infty}} + \frac{1}{k_2[A]}$$

where k_{exp} = the experimental rate constant

$$k_{\infty} = \frac{k_1 k_2}{k_{-2}}$$

$\frac{1}{[A]}$ could be expressed by $\frac{1}{P}$ because the pressure is proportional to [A].

CHAPTER 17

GRAVITATIONAL, ELECTRICAL AND MAGNETIC WORK

17.1 WORK AND ENERGY

Work is always given as the product of an intensive variable - one independent of the mass of the system - and an associated extensive variable.

$$dw = -\Phi dx$$

where Φ is the intensive variable and x is the extensive.

Some corresponding intensive and extensive variables are summarized as follows.

Type of Work	Φ_i (intensive)	X_i (extensive)
Thermal	T	S
Expansive	P	V
Gravitational	gh	M
Chemical	μ	n
Surface	γ	a

From the first law of thermodynamics

$$dE = dq + dw$$

where E is internal energy
q is heat added and
w is work.

In its general form, accounting for all work terms, the first law can be written

$$dE = TdS - PdV - \sum_i \Phi_i dx_i$$

If chemical work is included

$$dE = TdS - PdV - \sum_i \Phi_i dx_i + \sum_j \mu_j dn_j$$

where μ_j is the chemical potential of the jth species and n_j is the number of moles of the j species.

By taking suitable Legendre transforms of the energy function, the enthalpy, Helmholtz, and Gibbs energy functions can be derived for the augmented system.

Given

$$dE = TdS - PdV - \sum_i \Phi_i dx_i$$

and taking the transform with respect to work terms only, we derive the expression for enthalpy

$$H = E + PV + \sum_i \Phi_i x_i$$

$$dH = TdS + VdP + \sum_i x_i d\Phi_i$$

Now taking the transform with respect to thermal energy terms gives

$$A = E - TS$$

$$dA = -PdV - SdT - \sum_i \Phi_i dx_i$$

which is the Helmholtz function. Finally, taking the transform with respect to all energy terms (thermal and work) produces the Gibbs function

$$G = E - TS + PV + \sum_i \Phi_i x_i$$

$$dG = -SdT + VdP + \sum_i x_i d\Phi_i$$

dG, or more commonly ΔG, is a measure of the maximum useful work obtainable when a process is carried out under conditions of constant T, P, and in the augmented formulation, Φ_i. The equilibrium condition at constant T, P, and Φ_i is that $\Delta G = 0$.

17.2 GRAVITATIONAL WORK

If g can be considered constant, then

$$W = mg \int_{o}^{h} dh$$

where m is the mass of the body in question and dh is the height differential.

The Gibbs free energy expression including gravitational work then becomes

$$dG = -SdT + VdP + Mgdh$$

where M is the molar mass and

gdh is the intensive differential $d\Phi$ from

$$dG = -SdT + VdP + \sum_i x_i d\Phi_i$$

If dT = 0 and g is constant at equilibrium

$$VdP + Mgdh = 0$$

$$\frac{dP}{dh} = \frac{-Mg}{V} = -\rho g$$

If we consider a column of an ideal gas in which temperature is independent of height and is fixed, the above expression becomes

$$\frac{dP}{dh} = -\frac{MPg}{RT}$$

Integrating the above expression

$$\int_{P_1}^{P_2} \frac{dP}{P} = -\frac{Mg}{RT} \int_{h_0}^{h} dh$$

$$\ln \frac{P_1}{P_2} = -\frac{Mg}{RT} (h - h_0)$$

This is the expression for pressure as a function of height at constant temperature.

17.3 ELECTRICAL WORK AND THE PIEZOELECTRIC EFFECT

If an electric field is applied to a dielectric material, polarization results. The change in internal energy is given by

$$dE = \hat{E} d\overline{P} = dw$$

where $\hat{E}$ is the electric field strength and $\overline{P}$ is the polarizability of the material.

$$dG = -SdT + VdP - \overline{P} d\hat{E}$$

is then the expression for free energy.

If the length of the dielectric can be changed by applying a force or tension, the work done on the body is then

$$dw = tdl$$

where t is tension and

l is length.

Combining the electrical and mechanical components of work gives the equation for a body that undergoes a tension in an electric field.

$$dE = TdS - PdV + \hat{E} d\overline{P} + tdl$$

$$dG = -SdT + VdP - \overline{P} d\hat{E} - ldt$$

If $dT = dP = 0$

$$dG = -\bar{P}d\hat{E} - l dt$$

and at equilibrium

$$\frac{d\hat{E}}{dt} = \frac{-l}{\bar{P}}$$

Since dG is exact, we can apply Euler's formula

$$\left(\frac{\partial l}{\partial \hat{E}}\right)_{T,P,t} = \left(\frac{\partial \bar{P}}{\partial t}\right)_{T,P,\hat{E}}$$

Materials for which these derivatives are not ϕ exhibit piezoelectric properties.

17.4 MAGNETIC WORK AND ADIABATIC COOLING

The work associated with the magnetization of a mole of substance is

$$W = \int \bar{H}\, dM$$

where $\bar{H}$ is the applied magnetic field and M is the molar magnetic susceptibility.

The paramagnetic susceptibility of many substances can be represented by Curie's law

$$\chi = \frac{C}{T}$$

Where T is the temperature in °K and C is the Curie constant.

The equations of energy, enthalpy and free energy become

$$dE = TdS - PdV + \bar{H}\, dM$$

$$dH = TdS + VdP - Md\bar{H}$$

$$dG = -SdT + VdP - Md\bar{H}$$

Heat capacity at constant pressure and magnetic field can be defined as

$$C_{P,\bar{H}} = \left(\frac{\partial H}{\partial T}\right)_{P,\bar{H}} = T\left(\frac{\partial S}{\partial T}\right)_{P,\bar{H}}$$

Consider entropy as a function of magnetic field and temperature at constant pressure

$$S = S(\bar{H},T)_P$$

$$dS = \left(\frac{\partial S}{\partial T}\right)_{P,\bar{H}} dT + \left(\frac{\partial S}{\partial \bar{H}}\right)_{P,T} d\bar{H}$$

If entropy is constant, $dS = 0$, and

$$\left(\frac{\partial S}{\partial \bar{H}}\right)_{P,T} = -\left(\frac{\partial S}{\partial T}\right)_{P,\bar{H}}\left(\frac{\partial T}{\partial \bar{H}}\right)_{P,S}$$

where

$$\left(\frac{\partial S}{\partial T}\right)_{P,\bar{H}} = \frac{C_{P,\bar{H}}}{T}$$

Also from the Gibbs equation

$$dG = -SdT + VdP - Md\bar{H}$$

At constant P and making use of the Euler theorem for exact differentials

$$\left(\frac{\partial S}{\partial \bar{H}}\right)_{P,T} = \left(\frac{\partial M}{\partial T}\right)_{P,\bar{H}}$$

so that

$$\left(\frac{\partial T}{\partial \bar{H}}\right)_{P,S} = \frac{-T}{C_{P,\bar{H}}}\left(\frac{\partial M}{\partial T}\right)_{P,\bar{H}}$$

For a paramagnetic salt that obeys Curie's Law, $\partial M/\partial T$

is negative, so that $\partial T/\partial \bar{H}$ is positive. That is, the temperature increase with increasing magnetic field for an isentropic (dS = 0) process.

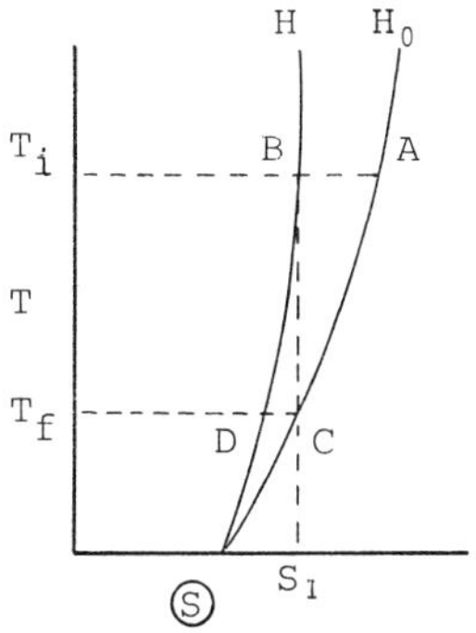

Fig. 17.1

If a paramagnetic salt sample is initially at A and is subjected to an applied magnetic field $\bar{H}$, the entropy decreases isothermally. When the field is removed reversibly while the sample is adiabatically isolated, the sample cools to T_f isentropically.

Since

$$M = \chi \bar{H}$$

$$dM = \chi d\bar{H} + \bar{H} d\chi$$

$$\left(\frac{\partial M}{\partial T}\right)_{P,\bar{H}} = \bar{H}\left(\frac{\partial \chi}{\partial T}\right)_{P,\bar{H}}$$

$$\chi = \frac{C}{T}; \quad \text{C is the Curie constant.}$$

$$\left(\frac{\partial \chi}{\partial T}\right)_{P,\bar{H}} = \frac{-C}{T^2}$$

$$\left(\frac{\partial M}{\partial T}\right)_{P,\bar{H}} = -\bar{H}\frac{C}{T^2}$$

so that

$$\left(\frac{\partial T}{\partial \bar{H}}\right)_{P,S} = -\frac{T}{C_{P,\bar{H}}}\left(\frac{\partial M}{\partial T}\right)_{P,\bar{H}} = \bar{H}\frac{C}{C_{P,\bar{H}}T}$$

$$dT = \frac{C}{C_{P,\bar{H}}T}\ \bar{H}\, d\bar{H}$$

An expression for the heat capacity can be found by taking the derivative of $C_{P,\bar{H}}$ with respect to $\bar{H}$.

$$C_{P,\bar{H}} = T\left(\frac{\partial S}{\partial T}\right)_{P,\bar{H}}$$

$$\frac{\partial}{\partial \bar{H}}(C_{P,\bar{H}})_T = T\,\frac{\partial^2 S}{\partial \bar{H}\,\partial T} = T\,\frac{\partial}{\partial T}\left(\frac{\partial S}{\partial \bar{H}}\right)_{P,T}$$

$$= T\,\frac{\partial}{\partial T}\left(\frac{\partial M}{\partial T}\right)_{P,\bar{H}} = T\,\frac{\partial^2 M}{\partial T^2}$$

$$= T\bar{H}\,\frac{\partial^2 \chi}{\partial T^2} \qquad \text{from } M = \chi\bar{H}$$

$$\chi = \frac{C}{T}$$

$$\frac{\partial^2 \chi}{\partial T^2} = \frac{2C}{T^3}$$

$$\frac{\partial C_{P,\bar{H}}}{\partial \bar{H}} = \frac{2C}{T^2}\,\bar{H}$$

$$C_{P,\bar{H}} = C_{P,\bar{H}}(T,O) + \frac{C}{T^2}\,\bar{H}^2$$

where $C_{P,\bar{H}}(T,O)$ is the heat capacity at zero magnetic field strength. This is approximately B/T^2 where B is a constant.

$$C_{P,\bar{H}} = \frac{B}{T^2} + \frac{C}{T^2}\,\bar{H}^2$$

Substituting this expression in the equation

$$dT = \frac{C}{C_{P,\bar{H}}T}\ \bar{H}\, d\bar{H}$$

gives

$$\frac{dT}{T} = \frac{C}{(B + C\bar{H}^2)}\ \bar{H}\, d\bar{H}$$

$$\ln\left(\frac{T_f}{T_i}\right) = \tfrac{1}{2}\ln\left(\frac{B + C\bar{H}_f^2}{B + C\bar{H}_i^2}\right)$$

where T_i, T_f, $\bar{H}_i$, $\bar{H}_f$ are the initial and final values of temperature and magnetic field respectively.

If $\bar{H}$ is significantly high enough so that B can be neglected, the final formula simplifies to

$$\frac{T_f}{T_i} = \frac{\bar{H}_f}{\bar{H}_i}$$

CHAPTER 18

SURFACE WORK

18.1 SURFACE TENSION AND SURFACE ENERGY

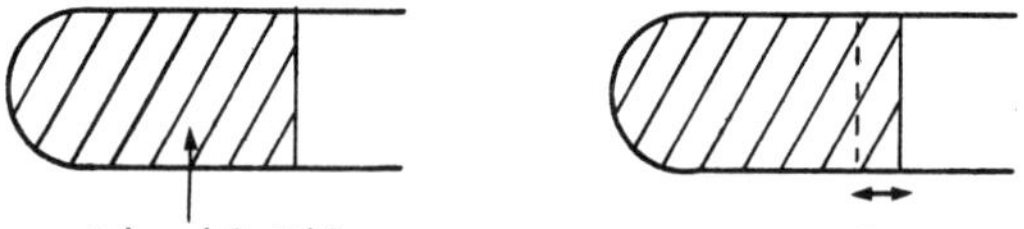

Fig. 18-1

For a liquid film in a wire frame work must be done to extend the frame beyond its equilibrium surface area. The force required to extend the film is proportional to 'L', the length of wire in contact with it, and γ the surface tension of the medium (N/m)

$$f = 2\gamma L$$

There is a factor of two because the upper and lower surfaces of the film define a boundary of 2L in contact with the slide wire.

$$dw = fdx = Z\gamma Ldx = \gamma da$$

Is the work exerted on the film, where in the last equation da = ZLdn is the total change in surface area?

$$dG = -SdT + VdP - ad\gamma$$

For equilibrium, the Helmholtz free energy must be minimized.

$$dA = -SdT - PdV + \gamma da = \phi$$

18.2 BUBBLES AND DROPS

If the pressure inside a sphere of liquid is P_{in}, the total outward force is $P_{in} 4\pi r^2$. The total inward force is P_{out}, the outside pressure and the surface tension.

$$4\pi r^2 \gamma = \gamma a = \text{Surface energy}$$

$$\gamma da = \text{Surface work} = 8\pi r \gamma\, dr$$

At mechanical equilibrium the forces must be equal.

$$P_{in} 4\pi r^2 = P_{out} 4\pi r^2 + 8\pi r \gamma$$

$$P_{in} - P_{out} = \frac{2\gamma}{r} \qquad r = \text{Radius of drop}$$

This is the Laplace equation. A more general form is

$$P_{in} - P_{out} = \gamma \left(\frac{1}{r_1} + \frac{1}{r_2} \right)$$

where r_1 and r_2 are two characteristic radii for a particular curved surface. If the surface is an air bubble the equation becomes

$$P_{in} - P_{out} = \frac{4\gamma}{r}$$

To account for the surface tension exerted by inner and outer sides of the bubble.

A droplet forming from a tube of radius 'r'. At the instant of breaking, the weight of the drop equals the surface force.

$$mg = 2\pi r \gamma \qquad [r = \text{Radius of capillary tube}]$$

$$\gamma = \frac{mg}{2\pi r}$$

This equation provides a direct way of measuring surface tension from drop weight and capillary.

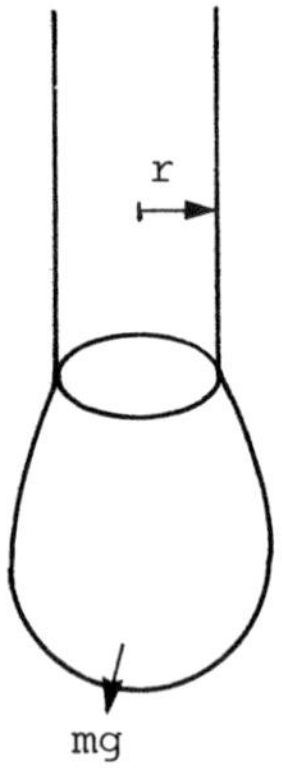

Fig. 18-2

Capillarity

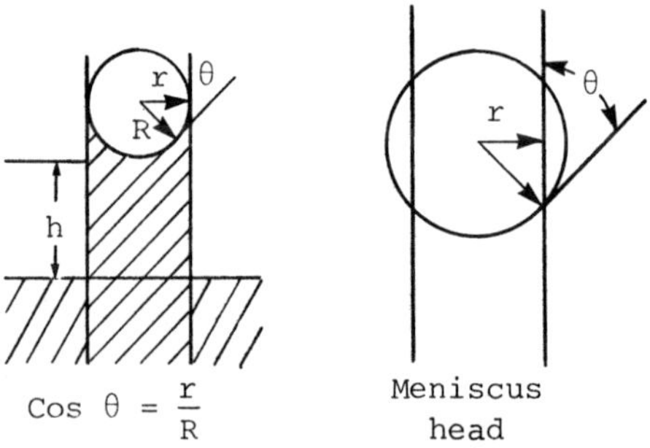

Fig. 18-3

θ = Contact angle

R = Radius of sphere approximating radius of curvature of meniscus

r = Radius of capillary.

$$(P'' - P') = \frac{2\gamma}{R} = \frac{2\gamma \cos\theta}{r}$$

If 'h' is the capillary height, the force per unit area of the cylindrical liquid column is $gh(\rho - \rho_0)$, where

ρ = density of liquid

ρ_0 = density of gas above it

$$(P'' - P') = \frac{2\gamma}{R} = \frac{2\gamma\cos\theta}{r} = gh(\rho - \rho_0)$$

$$\gamma = \tfrac{1}{2}gh(\rho - \rho_0)\frac{r}{\cos\theta}$$

Another way of approaching the problem is to sum the forces in equilibrium to ϕ.

The downward force of air pressure is $-P\pi r^2$, where 'r' is again the capillary radius. The upward force of pressure at depth 'd' in the trough is $(P + \text{rho})\pi r^2$. The downward force of gravity on the column is $-(h + d)\rho g\pi r^2$. An additional force due to surface tension, which acts tangent to the surface of the meniscus at an angle ϕ, has a vertical component $2\pi r\gamma\cos\phi$. Summing the respective forces to ϕ yields

$$(P + \rho gd)\pi r^2 - P\pi r^2 - (h + d)\rho g\pi r^2 + 2\pi r\gamma\cos\phi$$

$$= \phi$$

which simplifies to

$$\frac{\rho ghr}{2\cos\phi} = \gamma$$

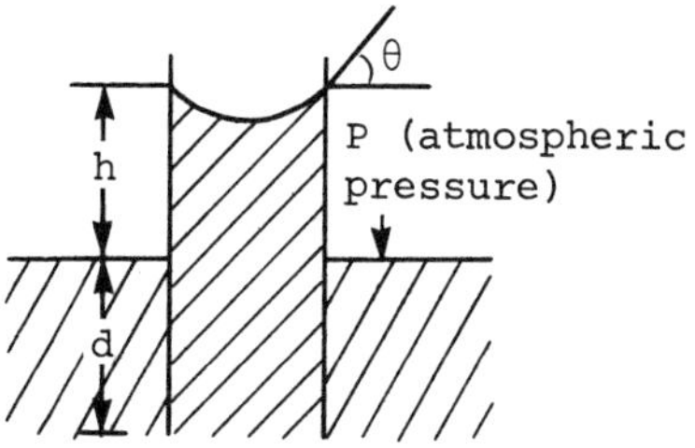

Fig. 18-4

18.3 THE KELVIN EQUATION

Consider a spherical drop of pure liquid with radius of curvature $1/r$ and surface tension γ equilibrium with its vapor.

$$(P'' - P') = \frac{2\gamma}{r} \quad \text{where} \quad P'' = P_{internal}$$

$$P' = P_{external}$$

For mechanical equilibrium

$$dP'' - dP' = d\left(\frac{2\gamma}{r}\right)$$

At physiochemical equilibrium for two phases

$$\mu_i' = \mu_i'' \quad \text{and} \quad d\mu_i' = d\mu_i''$$

$$d\mu_i = -S_i dT + V_i dP$$

At constant 'T'

$$V_i' dP' = V_i'' dP''$$

and the equation for mechanical equilibrium becomes

$$d\left(\frac{2\gamma}{r}\right) = \left(\frac{V_i' - V_i''}{V_i''}\right) dP'$$

Neglecting the molar volume of the liquid in comparison to the gas, since it is so much smaller, and treating the vapor as an ideal gas gives

$$V_i' = RT/P'$$

$$d\left(\frac{2\gamma}{r}\right) = \frac{RT}{V_i''} \cdot \frac{dP'}{P'}$$

Integrating over the appropriate limits

$$\int_0^{1/r} d\left(\frac{2\gamma}{r}\right) = \int_{P_0 = V.P.}^{P} \frac{RT}{V_i''} \cdot \frac{dP'}{P'}$$

$$\ln\frac{P}{P_0} = \frac{2\gamma}{r} \cdot \frac{V_i''}{RT}$$

If $V_i'' = M/\rho$ where M = Molar mass

ρ = Density of pure fluid

$$\ln \frac{P}{P_0} = \frac{2\gamma}{r} \cdot \frac{M}{RT\rho}$$

18.4 GIBBS FORMULATION OF ADSORPTION

Consider the change in free energy at the interface of a two phase system.

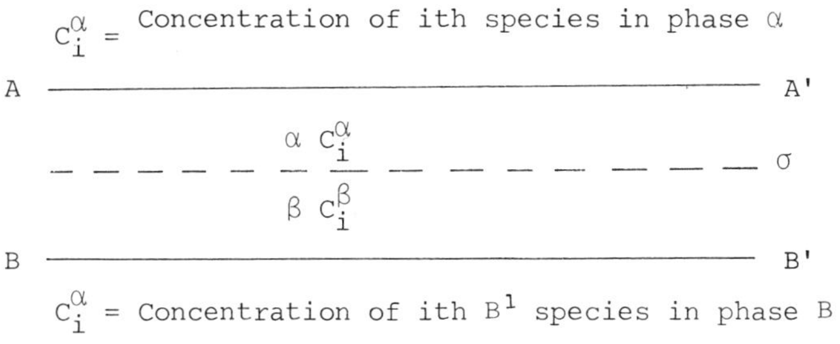

Fig. 18-5

$$dG^{\sigma} = -S^{\sigma}dT + \gamma\, dA + \sum_i \mu_i dn_i^{\sigma}$$

where the superscript 'σ' denotes the interfacial region. At equilibrium

$$\mu_i^{\sigma} = \mu_i^{\alpha} = \mu_i^{\beta}$$

where μ_i^{α} = Molar potentials ith species phase 'A'
μ_i^{β} = Molar potentials ith species phase 'B'

and at constant 'T'

$$dG^{\sigma} = \gamma\, dA + \Sigma \mu_i dn_i$$

Integrating at constant γ and μ_i gives

$$G^{\sigma} = \gamma A + \Sigma \mu_i dn_i^{\sigma}$$

from which it is apparent after differentiating and comparing to the previous equation

$$\Sigma n_i^{\sigma} d\mu_i + A d\gamma = \phi$$

$$d\gamma = -\Sigma \frac{n_i^{\sigma}}{A} d\mu_i = -\Sigma \Gamma_i d\mu_i$$

where

$$\Gamma_i = \frac{n_i^{\sigma}}{A}$$

The explicit form of this equation for a two component solution is

$$d\gamma = -\Gamma_i d\mu_i - \Gamma_2 d\mu_2$$

Since μ_1 and μ_2 cannot be independently valued we must use the concept of relative adsorptions. To do this we choose an interfacial region, σ, between α and β where the adsorption of component 1 is ϕ. The relative adsorption is then denoted

$$\Gamma_{i,1} \quad \text{i.e.} \quad \Gamma_{i,1} = \phi$$

$$\Gamma_{2,1} = -\left(\frac{\partial\gamma}{\partial\mu_2}\right)_T = -\frac{1}{RT}\left(\frac{\partial\gamma}{\partial \ln C_2}\right)_T$$

For $\mu_2 = \mu_2^{\theta} + RT\ln C_2$

18.5 THE LANGMUIR ADSORPTION ISOTHERM

If the rate of adsorption can be assumed to be proportional to the number of unoccupied sites on an adsorbing surface and the pressure above the surface, neglecting energies of interaction at occupied sites, it can be given as

$$ka(M - N)P$$

where M = Total number of sites

N = Total number of occupied sites

P = Pressure

ka = Rate of adsorption.

In the same way the rate of deadsorption is proportional to the number of occupied sites, kdN.

At equilibrium the two rates are equal

$$ka(M - N)P = kdN$$

$$ka(1 - \theta)P = kd\theta \qquad \theta = N/M$$

$$\theta = \frac{P\kappa}{1 + P\kappa}$$

where 'κ' is the equilibrium constant. This is the Langmuir adsorption isotherm.

18.6 STATISTICAL MECHANICS OF ADSORPTION

The Langmuir isotherm can be deduced from a statistical mechanical approach to the problem. The partition function for a one-dimensional lattice with 'M' sites and 'N' adsorbed molecules is

$$Z(N,M,T) = z^n \sum_{Q+1} g(M,N,Q)e^{-[N-Q+\frac{1}{2})^{w/kt}}$$

where z = Partition function for a single molecule

w = Surface interaction energy

Q = Number of partitions between occupied and unoccupied sites.

$g(M,N,Q)$ = Statistical weight of the system, i.e. the number of permutations with the same total energy of interaction k = Boltzman constant

T = (Temperature) °K

M = Total number of unoccupied sites

N = Total number of occupied sites.

The total energy of interaction is a function of the number of sites with adjacent molecular adsorption. Consider the following model

XX	OO	X	O	XXX	O

where O = empty, X = occupied.

If Q is the number of OX-pairs then

The total energy of interaction $= N_{XX}w = \left(N - \frac{Q+1}{2}\right)w = 3w$

and N_{XX} is the number of groups of X-sites.

For the above model

$$N_{XX} = 3$$
$$Q = 5$$
$$N = 6$$

In the Langmuir approach the surface interaction energy 'W' is ϕ The partition function then simplifies to

$$Z(N,M,T) = z^n \sum_Q g(M,N,Q)$$

The number of ways 'M' spaces can be filled by 'N' items is given by the expression

$$\frac{M!}{N!(M-N)!} = g(M,N,Q)$$

The partition function then becomes

$$Z(N,M,T) = z^n \frac{M!}{N!(M-N)!}$$

For the surface phase the Gibbs free energy is given by

$$G^{\sigma} = -kT \ln Z^{\sigma}$$

and the general expression is

$$dG^{\sigma} = -S^{\sigma} dT + \gamma dA + \Sigma \mu dN$$

where 'μ' is the chemical potential per molecule.

Since it can be expressed as

$$dA = \alpha dM$$

where 'M' is the number of sites and 'α' is a proportionality factor. We have at equilibrium

$$\left(\frac{\partial G^{\sigma}}{\partial N}\right)_{M,T} = \mu = -kT\left(\frac{\partial \ln Z^{\sigma}}{N}\right)_{M,T}$$

Using Stirling's approximation

$$N! = N \ln N - N$$

$$\frac{\mu}{kT} = -\left(\frac{\partial \ln Z}{\partial N}\right)_{M,T} = \ln \frac{N}{z(M - N)}$$

$$\frac{N}{M} = \theta$$

$$\ln \frac{N}{z(M - N)} = \ln \frac{\theta}{z(1 - Q)} = \frac{\mu}{kT}$$

At equilibrium with a gas at constant P

$$\mu = \mu^0 + kT \ln P$$

$$\frac{\mu}{kT} = \frac{\mu^0}{kT} + \ln P = \ln \frac{\theta}{z(1 - \theta)}$$

$$Pe^{\mu^0/kT} = \frac{\theta}{z(1 - \theta)}$$

If $ze^{\mu^0/kT} = b(T)$

$$\frac{\theta}{(1 - \theta)} = Pb(T)$$

which reduces to the Langmuir isotherm

$$\theta = \frac{Pb(T)}{Pb(T) + 1}$$

where k_{eq} is now some function of T and the energy of individual molecules.

18.7 THE ISING ADSORPTION MODEL

If the interaction energy is not zero the evaluation of g(M,N,Q) bcomes complete especially when M, N and Q are the order of one mole (6.02×10^{23}).

If 'N' is the number of adsorbed species and 'Q' is again the number of boundaries between adsorbed and non-adsorbed groups, then the number of arrangements of groups from 'N' molecules is given by

$$\frac{N!}{(Q/2)!(N - Q/2)!}$$

The actual number of groups adsorbed is $\left(\frac{Q + 1}{2}\right)$. However 'Q' in most cases is so large that Q/2 suffices. The number of arrangements of unoccupied sites becomes

$$\frac{(M - N)!}{(Q/2)!(M - N - Q/2)!}$$

The statistical function g(M,N,Q) is then equal to twice the product of these two distributions, since any combination could be written backwards as well as forwards.

The partition function becomes

$$Z(N,M,T) = (ze^{-w/kt})^N \sum_Q t^*(N,M,Q)$$

$$t^*(N,M,Q) = g^*(e^{w/2kt})^Q$$

t* is the value of t(N,M,Q) for which Q = Q*, when Q = Q* the statistical function 'g' is at a maximum, given by g*

$$g^*(N,M,Q) = Z\begin{pmatrix} N \\ \left(\frac{Q}{Z}\right) \end{pmatrix}\begin{pmatrix} (M-N) \\ \left(\frac{Q}{Z}\right) \end{pmatrix}$$

$$g^*(N,M,Q) = \frac{ZN!(M-N)!}{(N-Q/Z)!(M-N-Q/Z)![(Q/2)!]^2}$$

To evaluate t* we proceed as follows

$$\frac{d\ln t}{dQ} = ¢ = \frac{d\ln g}{dQ} + \frac{w}{2kt}$$

Applying the Stirling approximation to 'g' and differentiating with respect to Q gives

$$\ln(N - Q/2) + \ln(M - N - Q/2) - 2\ln Q/2 = -w/kt$$

or $$\frac{(N - Q/2)(M - N - Q/2)}{(Q/2)^2} = e^{-w/kT}$$

$$\frac{(\theta - y)(1 - \theta - y)}{y^2} = e^{-w/kT}$$

where $\theta = N/M$ and $y = Q/2M$.

Using the quadratic formula and algebraic manipulations yields for the greater of the two roots.

$$y = \frac{2\theta(1-\theta)}{(1+\beta)} \qquad \beta = \sqrt{1 - 4\theta(1-\theta)(1-e^{-w/2kt})}$$

We now have a value of Q* in terms of 'W' and 'N'. Evaluating the natural log of the partition function gives

$$\ln Z = N\ln(ze^{-w/kt}) + \ln t^*$$

and from

$$\mu = -kT\left(\frac{\partial \ln Z}{\partial N}\right)_{M,T}$$

$$\frac{-\mu}{kT} = \left(\frac{\partial \ln Z}{\partial N}\right)_{M,T} = \ln(ze^{-w/kt}) + \left(\frac{\partial \ln t^*}{\partial N}\right)_{M,T}$$

$$\frac{-\mu}{kT} = \left(\frac{\partial \ln Z}{\partial N}\right)_{M,T} = \ln(ze^{-w/kt}) + \left(\frac{\partial \ln g^*}{\partial N}\right)_{M,T}$$

where g* is the maximum value obtained by using Q*, and Q*W/2kt is small enough in comparison to be ignored. Then

$$\left(\frac{\partial \ln g^*}{\partial N}\right)_{M,T} = \ln \frac{N(M - N - Q/2^*)}{(M - N)(N - Q/2)}$$

$$\frac{-\mu}{kT} = \ln ze^{-w/kT} + \ln \frac{\theta(1 - \theta - y)}{(1 - \theta)(\theta - y)}$$

From $\qquad N/M = \theta \quad \text{and} \quad \frac{Q}{2M} = y$

$$e^{-\mu/kT} = (ze^{-w/kT}) \cdot \left(\frac{\theta(1 - \theta - y)}{(1 - \theta)(\theta - y)}\right)$$

Introducing the value of 'y' for which Q* is maximum yields

$$y = \frac{2\theta(1 - \theta)}{1 + \beta}$$

$$\lambda ze^{-w/kT} = \frac{\beta - 1 + 2\theta}{\beta + 1 - 2\theta}; \quad \text{where} \quad \lambda = e^{\mu/kT}$$

For a gas in equilibrium

$$\mu = \mu^0 + kT \ln P$$

$$e^{\mu/kT} = Pe^{\mu^0/kT} = \lambda$$

and finally the adsorption equation becomes

$$Pze^{\frac{\mu^0 - w}{kT}} = \frac{\beta - 1 + 2\theta}{\beta + 1 - 2\theta}$$

CHAPTER 19

KINETIC THEORY

19.1 STATISTICAL METHODS

The probability of an event is the number of possible occurrences of that event divided by the total number of possible outcomes. Independent probabilities are multiple in nature for instance:

P(h,h) is the probability of getting two heads when flipping two coins, where $P(h) = P(t) = \frac{1}{2}$. Since the two events are independent: $\frac{1}{2} \cdot \frac{1}{2} = \frac{1}{4}$

$$\begin{aligned} P(h,h) &= \tfrac{1}{4} \\ P(h,t) &= \tfrac{1}{4} \\ P(t,h) &= \tfrac{1}{4} \\ P(t,t) &= \underline{\tfrac{1}{4}} \\ & \quad 1 \end{aligned}$$

The probability then of getting a head and a tail is $\frac{1}{4}$. probability of getting a head and a tail irrespective of order is $\frac{1}{2}$, since there is twice as much a chance of getting P(h,t) or P(t,h), as there is of getting either P(h,h) or P(t,t).

The sum of all the probabilities of individual outcomes must sum to one.

19.2 THE BINOMIAL DISTRIBUTION

If P is the probability of an event and $(1 - P) = q$

is the probability that the event does not occur, then in 'N' trials with 'n' successful attempts, with each trial being independent,

$$P'(n,(N - n)) = p^n q^{(N-n)}$$

Since the 'n' successes and (N - n) failures can occur in any one of $\frac{N!}{n!(N - n)!}$ total possible combinations we get

$$P(n,(N - n)) = \frac{N!}{n!(N - n)!} \quad p^n q^{(N - n)}$$

This is the binomial distribution, also written as

$$\sum_{n=0}^{N} \binom{N}{n} p^n q^{(N-n)} = q^N + \binom{N}{1} p\, q^{(N-1)} + \left[\frac{N}{2}\right] p^2 q^{(N-2)} + \ldots + \binom{N}{n} p^n q^{N-n} + \ldots + p^N$$

$$= 1$$

As $N \to \infty$, the probability of achieving a distribution varying much from the most probable distribution, the binomial distribution, is vanishingly small.

19.3 THE GAUSSIAN DISTRIBUTION

If we take the average of the outcomes of 'n' experiments, say 'n' randomly generated numbers between 0 and 1, over N trials we can plot a frequency distribution known as the Gaussian distribution for the event, which represents the frequency of the averages. The formula for the Gaussian distribution is

$$f(x) = \frac{h}{\sqrt{\pi}} \quad e^{[-h^2(x-m)^2]}$$

where m is the mean of the trials, h is $\frac{1}{\sqrt{2}\sigma}$, σ is the standard deviation, and σ^2 is known as the variance

$$\sigma^2 = \int_{-\infty}^{\infty} (x - m)^2 f(x)dx = \frac{1}{2h^2}$$

The probability of an outcome occuring between a and b on the Gauss curve is

$$P(a,b) = \int_a^b f(x)dx$$

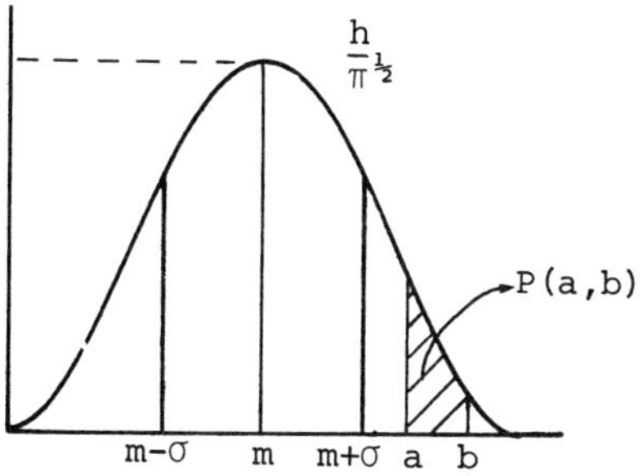

Fig. 19-1

The area under the curve between m − σ and m + σ is .683 of the unit area. Or, put another way, 68.3% of the outcomes of a given event occur within ± σ, one standard deviation, of the mean. The Gauss function is more commonly written

$$I = \frac{1}{\sqrt{2\pi}} \int_o^z e^{-\frac{1}{2}z^2} dz$$

where $z = h(x - m)$

The variance of a small number of measurements is given by

$$\sigma^2 = \frac{1}{N} \sum_i (x_i - \bar{x})^2$$

where x_i is the ith measurement, $\bar{x}$ is the average of measurements, and n is the number of measurements.

19.4 THE BOLTZMANN DISTRIBUTION

Given a collection of N molecules having total energy E and distributed among ε_i discrete energy levels, the fraction of molecules with energy ε_i compared to N_0 with ε_0 is given by the Boltzmann distribution law

$$N_i = N_o e^{-(\varepsilon_i - \varepsilon_o)/kT}$$

where k is the Boltzmann constant and T is the temperature in degrees kelvin.

The law gives the ratios of numbers of molecules with different energies as a function of ε and T, $\varepsilon_i - \varepsilon_o$ can also be written $\Delta\varepsilon$.

$$N_i = N_o e^{-\Delta\varepsilon/kT}$$

If the energy is sealed so that Eo = 0 we have

$$Ni = N_o e - Ei/kT$$

19.5 ONE-DIMENSIONAL VELOCITY DISTRIBUTION

The kinetic energy of a molecule with mass 'm' and velocity 'μ' moving along x-axis is given by $\frac{1}{2}m\mu^2$. Making the appropriate substitution into the Boltzmann equation, and with $\mu = 0$ at $E = 0$, the number of particles with a velocity between μ and $\mu + d\mu$ becomes

$$dN = dN_o e^{-m\mu^2/2kT}$$

where dN is the number of particles with velocities between μ and $\mu + d\mu$ and dN_o is the number of particles with velocities between 0 and $d\mu$.

dN_o is proportional to $d\mu$, and substituting, $\alpha\, d\mu$ for dN_o and dividing by 'N' gives the number of particles in 'N' with velocities between $\mu + d\mu$ and μ.

$$\frac{dN}{N} = \frac{\alpha}{N} e^{-\frac{m\mu^2}{2kT}} d\mu = f(\mu)d\mu$$

Normalization of the above equation yields for the constant $\frac{\alpha}{N} = A$

$$A \int_{-\infty}^{\infty} e^{-\frac{m\mu^2}{2kT}} d\mu = 1$$

with
$$s^2 = \frac{m\mu^2}{2kT}$$

$$ds = \left(\frac{m}{2kT}\right)^{\frac{1}{2}} d\mu$$

and
$$A \left(\frac{2kT}{m}\right)^{\frac{1}{2}} \int_{-\infty}^{\infty} e^{-s^2} ds = 1$$

The integral in terms of s is a standard one and its value is $\pi^{\frac{1}{2}}$. Therefore

$$A = \left(\frac{m}{2\pi kT}\right)^{\frac{1}{2}}$$

By comparison with the Gauss function,

$$h = \left(\frac{m}{2kT}\right)^{\frac{1}{2}}$$

For the distribution and the mean of μ is zero. The fraction of molecules with velocities between μ_A and μ_B is then found by

$$\int_{\mu_A}^{\mu_B} f(\mu)d\mu$$

To evaluate the mean of μ we use the averaging technique for a continuous distribution

$$\bar{\mu} = \frac{\int_{-\infty}^{\infty} f(\mu)d\mu}{\int_{-\infty}^{\infty} f(\mu)d\mu}$$

Since the denominator in the last expression is unity, this becomes

$$\bar{\mu} = \int_{-\infty}^{\infty} \left(\frac{m}{2\pi kT}\right)^{\frac{1}{2}} e^{-\frac{m\mu^2}{2kT}} \mu\, d\mu$$

A standard table of integrals will show

$$\int_{-\infty}^{\infty} xe^{-ax^2} dx = 0$$

and therefore

$$\bar{\mu} = 0$$

The root mean square velocity can be found by a similar procedure

$$(\bar{\mu}^2) = \int_{-\infty}^{\infty} \mu^2 f(\mu)d\mu$$

Again, a standard table of integrals will show that

$$\int_{-\infty}^{\infty} x^2 e^{-ax^2} dx = 2\int_{0}^{\infty} x^2 e^{-ax^2} dx = \frac{\sqrt{\pi}}{2} a^{-\frac{3}{2}}$$

and with

$$a = \frac{m}{2kT}$$

$$(\bar{\mu}^2) = \left(\frac{m}{2\pi kT}\right)^{\frac{1}{2}} \cdot \pi^{\frac{1}{2}} \cdot \left(\frac{2kT}{m}\right)^{\frac{3}{2}}$$

$$(\bar{\mu}^2) = \frac{kT}{m}$$

The root mean square speed is then

$$(\bar{\mu}^2)^{\frac{1}{2}} = \sqrt{\frac{kT}{m}}$$

19.6 SOME USEFUL INTEGRALS

1) $$\int_0^{\infty} x^{2n} e^{(-ax^2)} dx = \frac{\sqrt{\pi}}{2} \frac{(2n)!a - (n + \frac{1}{2})}{2^{2n} n!}$$

2) $$\int_{-\infty}^{\infty} x^{2n} e^{(-ax^2)} dx = 2 \int_0^{\infty} x^{2n} e^{(-ax^2)} dx$$

3) $$\int_{-\infty}^{\infty} x^{(2n+1)} e^{(-ax^2)} dx = 0$$

4) $$\int_0^{\infty} x^{(2n+1)} e^{(-ax^2)} dx = \frac{\frac{1}{2} n!}{a^{(n+1)}}$$

19.7 HIGHER DIMENSIONAL DISTRIBUTIONS

A two-dimensional interpretation of the molecular velocity problem requires that the X and Y velocity components be distributed along a Gaussian surface oriented about the X and Y axes. The volume under the surface is unity, and the probability of a molecule having simultaneous velocity components between μ_x and $\mu_x + d\mu_x$ and μ_y and $\mu_y + d\mu_y$ is the volume bounded by the Gaussian surface and the $d\mu_x$ $d\mu_y$ region below it. It is more convenient to convert the two-dimensional problem into a one-dimensional one via the transform.

$$\mu_{xy} = (\mu_x^2 + \mu_y^2)^{\frac{1}{2}}$$

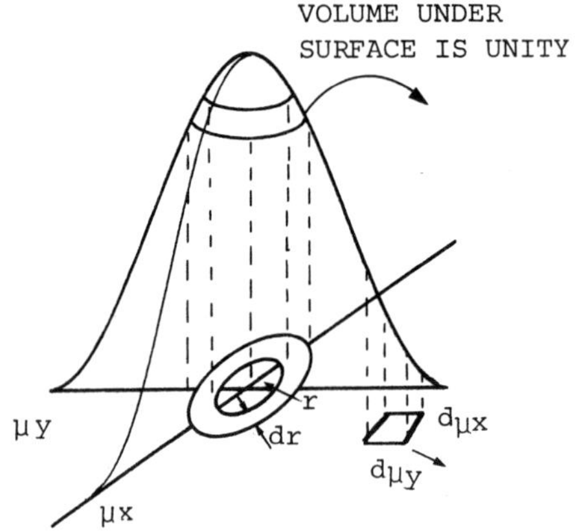

Fig. 19-2 Parallelepied of probability element for velocity μxy between μx and μx + dμx and μy and μy + dμy.

Since the probabilities for μ_x and μ_y values are independent of each other, the Gaussian distribution becomes

$$f(\mu_x \cdot \mu_y) = \frac{m}{2\pi kT} e^{[-\frac{m}{2kT}(\mu_x^2 + \mu_y^2)]}$$

Changing to polar coordinates with

$$\mu_x = r\cos\theta$$

$$\mu_y = r\sin\theta$$

$$\mu_{xy} = r$$

gives

$$f(r, \theta)rdrd\theta = \frac{m}{2\pi kT} e^{-\frac{m}{2kT}r^2} rdrd\theta$$

The angular dependence is symmetric about the origin; we therefore can eliminate it by integrating θ from 0 to 2π.

$$f(r)rdr = \frac{m}{kT} e^{-\frac{m}{2kT} r^2} rdr$$

In the same way, the three-dimensional distribution

$$f(\mu_x, \mu_y, \mu_z) = \left(\frac{m}{2\pi kT}\right)^{\frac{3}{2}} e^{[-\frac{m}{2kT}(\mu_x^2 + \mu_y^2 + \mu_z^2)]}$$

can be transformed into a one-dimensional distribution by converting to spherical coordinates and integrating over the symmetric angular regions. $(0 \leq \theta \leq 2\pi;\ 0 \leq \phi \leq \pi)$

$$d\mu_x d\mu_y d\mu_z = dV = r^2 \sin\phi \, dr d\phi \, d\theta$$

$$r = (\mu_x^2 + \mu_y^2 + \mu_z^2)^{\frac{1}{2}}$$

$$\int_{\text{all space}} f(\mu_x, \mu_y, \mu_z) d\mu_x d\mu_y d\mu_z = \int_0^{2\pi} \int_0^{\pi} \int_0^{r} \left(\frac{m}{2\pi kT}\right)^{\frac{3}{2}} e^{-\frac{m}{2kT} r^2} r^2 \sin\phi \, dr d\phi \, d\theta$$

$$f(r) = 4\pi \left(\frac{m}{2\pi kT}\right)^{\frac{3}{2}} e^{-\frac{m}{2kT} r^2} r^2$$

Average speeds and root mean square speeds can now be found as in the one-dimensional case.

$$\overline{C} = \int_0^{\infty} Cf(c)dc = 4\pi \left(\frac{m}{2\pi kT}\right)^{\frac{3}{2}} \int_0^{\infty} e^{-\frac{mc^2}{2kT}} C^3 dc$$

$$\overline{C} = \left(\frac{8kT}{\pi m}\right)^{\frac{1}{2}}$$

'C' has been substituted for 'r' in the original distribution function, and the limits, 0 to ∞, are used because of the symmetry of the distribution (the molecular speed 'C', in a spherical region $4\pi c^2 dc$, is the same irrespective of angular orientation). The integral can be evaluated using the table in section 6. In addition

$$C_{RMS} = (\overline{C^2})^{\frac{1}{2}} = 4\pi \left(\frac{m}{2\pi kT}\right)^{\frac{3}{2}} \int_0^{\infty} e^{-\frac{mc^2}{2kT}} C^4 dc$$

$$= \left(\frac{3kT}{m}\right)^{\frac{1}{2}}$$

19.8 THE AVERAGE KINETIC ENERGY AND THE MOST PROBABLE SPEED

To obtain the average kinetic energy we multiply the observable, $\frac{1}{2}mc^2$, by the probability function and integrate

$$\overline{\tfrac{1}{2}mc^2} = \frac{m}{2}\int_0^\infty C^2 f(c)\,dc$$

$$= \frac{m}{2}\,4\pi\left(\frac{m}{2\pi kT}\right)^{\frac{3}{2}}\int_0^\infty e^{\frac{-mc^2}{2kT}}\,C^4\,dc$$

$$= 2\pi m\left(\frac{m}{2\pi kT}\right)^{\frac{3}{2}}\tfrac{1}{2}\sqrt{\pi}\,\frac{4!}{2^4\,2!}\left(\frac{2kT}{m}\right)^{\frac{3}{2}}$$

$$= \frac{3}{2}kT$$

which is in agreement with the equipartition of energy principle, allowing $\frac{1}{2}kT$ for each degree of freedom when translational motion alone is considered.

The most probable speed occurs at the maximum of the probability distribution curve. It is found by taking the first derivative of the probability function and setting it equal to zero.

$$f(c) = 4\pi\left(\frac{m}{2\pi kT}\right)^{\frac{3}{2}} e^{-\frac{mc^2}{2kT}}\,C^2$$

$$\frac{d}{dc}(f(c)) = f(c)\left\{2c - \frac{m}{2kT}2c^3\right\} = 0$$

Dividing throughout by like factors

$$C^2 = \frac{2kT}{m}$$

$$C_{mp} = \sqrt{\frac{2kT}{m}}$$

19.9 THE PRESSURE OF A GAS

The density function $f(\mu)d\mu$, gives a fraction of molecules with velocities between μ and $\mu + d\mu$

$$\frac{dN(\mu)}{N} = f(\mu)d\mu$$

In a time interval dt, those molecules with velocities between μ and $\mu + d\mu$ and contained in a volume Adt, where A is a small surface area element, will collide with the surface. The number of molecules colliding with the surface area 'A' is then

$$\frac{dN(\mu)}{N} A\mu dt = f(\mu)A\mu\, d\mu\, dt$$

which can also be written

$$\frac{dN(\mu)}{V} A\mu\ dt = \frac{N}{V} f(\mu)A\mu d\mu\ dt$$

where $\frac{dN(\mu)}{V}$ is the number of molecules per unit volume in the selected velocity range, μ and $\mu + d\mu$.

During collision with the wall of a vessel containing the molecules, the magnitude of the change in momentum per molecule is $2m\mu$. The total change in momentum is

$$dp = (2m\mu) \frac{NA}{V} \mu f(\mu)d\mu dt$$

Pressure is the force (rate of change of momentum) per unit area

$$dP = \frac{dp/dt}{A} = \frac{2Nm}{V} \mu^2 f(\mu)d\mu$$

We know from our previous development

$$\overline{U^2} = 2 \int_0^\infty \mu^2 f(\mu) d\mu$$

for the one-dimensional case. So

$$P = Nm \frac{\overline{U^2}}{V}$$

But

$$\overline{C^2} = \overline{\mu_x^2} + \overline{\mu_y^2} + \overline{\mu_z^2}$$

$$\overline{\mu_x^2} = \overline{\mu_y^2} = \overline{\mu_z^2}$$

so that

$$P = \frac{Nm}{3V} \overline{C^2}$$

for an ideal gas whose molecular constituents undergo elastic collisions.

CHAPTER 20

COLLISIONAL AND TRANSPORT PROPERTIES OF GASES

20.1 APPROXIMATE SOLUTION OF MOLECULAR EFFUSION

If we assume that 1/3 of all molecules are oriented along each axis, x, y, z, and that $\frac{1}{2}$ of each of those molecules are oriented in either + or - velocity directions, then the number of molecules striking the area element dS in a unit time is

$$\frac{dn}{dt} = \frac{1}{6}\, n^*\bar{c}dS$$

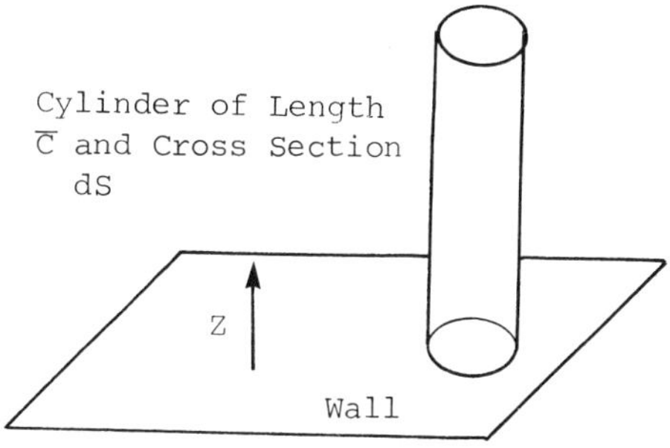

Fig. 20.1

where n* is the number of molecules per cubic centimeter; $\bar{c}$ is the average speed of the molecules, and the factor $\frac{1}{6}$ accounts for those molecules with proper orientation with respect to the chosen axis

$$\bar{c} = \left(\frac{8kT}{\pi m}\right)^{\frac{1}{2}}$$

$$\frac{dn}{dt} = \frac{1}{6}\, n^* \left(\frac{8kT}{\pi m}\right)^{\frac{1}{2}} dS$$

where m is the mass of the molecular constituent of the given gas (M/L°).

If we replace n*, the number of molecules per cm^3, by its identity from the ideal gas law

$$n^* = P/kT$$

$$\frac{dn}{dt} = \frac{1}{6} \left(\frac{8}{\pi mkT}\right)^{\frac{1}{2}} PdS$$

Since dn/dt is the number of molecules effusing per second, the weight loss per second can be written

$$-\frac{dw}{dt} = m\,\frac{dn}{dt} = \frac{P}{6}\left(\frac{8m}{\pi kT}\right)^{\frac{1}{2}} dS$$

This equation provides a way to measure vapor pressures by measuring the rate of weight loss in a sample. The equations derived thus far are true if no collisions between molecules occur near the orifice. In order to assure this we must make sure that the orifice is smaller in dimensions than the mean free path. The mean free path is the average distance a molecule travels before encountering another molecule.

For hydrogen molecules (M = 2) at 298°K and 1 atmosphere

$$n^* = \frac{P}{kT} = \frac{1 \text{ atm}}{\left(\frac{82.06}{6.02 \times 10^{23}} \;\frac{cm^3\ atm}{mol°K}\right)(298°K)}$$

$$= 2.462 \times 10^{19}\ \frac{molecules}{cm^3}$$

$$\bar{c} = \left(\frac{8kT}{\pi m}\right)^{\frac{1}{2}} = \left(\frac{(8)(1.38 \times 10^{-16}\ \frac{erg}{°K})(298°K)}{(3.1415)(2/6.02 \times 10^{23})}\right)^{\frac{1}{2}}$$

$$\bar{c} = 1.775 \times 10^5 \ \frac{cm}{S}$$

The number of collisions per cm^2 is then

$$\frac{dn}{dt} = \frac{1}{6} \ n^*\bar{c}dS$$

$$= \frac{(2.462 \times 10^{19} \frac{molecules}{cm^3})(1.775 \times 10^5 \frac{cm}{S})(cm^2)}{6}$$

$$= 7.28 \times 10^{23} \ \frac{molecules}{S}$$

20.2 EXACT SOLUTION OF MOLECULAR EFFUSION

In order to more accurately compute the number of molecules hitting a surface area element dS, we must consider the fraction of moelcules having the necessary speed and orientation. Consider a differential volume element containing molecules with velocity distributions given by f(c)

$$f(c) = 4\pi \left(\frac{m}{2\pi kT}\right)^{\frac{3}{2}} e^{\frac{-mc^2}{2kT}} \ c^2$$

The volume element is given as

$$dV = r^2 dr \ \sin\phi \, d\phi \ d\theta$$

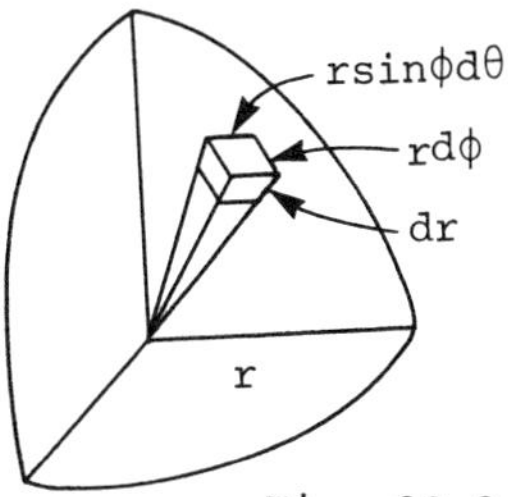

Fig. 20.2 (A)

Of the molecules in dV, only those traveling in the proper direction will ever hit dS. This fraction of molecules is contained in the solid angle subtended by dS at dV.

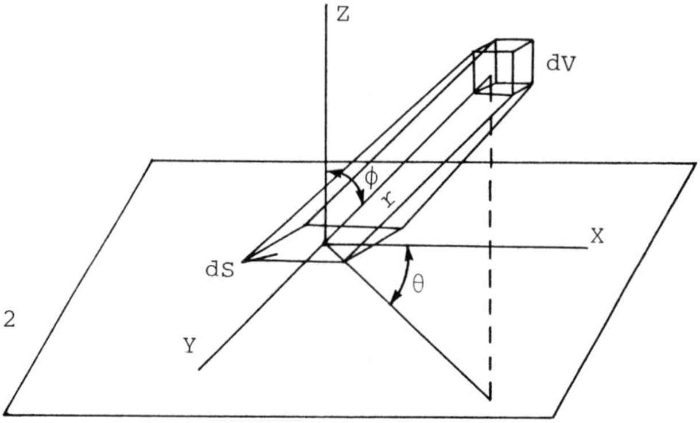

Fig. 20.2
(B)

A solid angle is defined as the ratio of an area of spherical surface to the square of its radius.

$$d\omega = \frac{dS}{r^2}$$

Hence, the total solid angle subtended by a sphere is $\omega = 4\pi$. The solid angle in question is equal to the ratio of the area $\cos\phi dS$, the projection of dS on a sphere of radius r, to the square of the radius. The fraction of the surface area covered by the solid angle is accordingly

$$\frac{d\omega}{4\pi} = \frac{\cos\phi\, dS}{4\pi r^2}$$

The number of molecules leaving dV with the proper orientation and velocities to collide with dS is then

$$\frac{dn}{dt} = n^* \; [f(c)dc][r^2 dr \sin\phi\, d\phi\, d\theta] \left[\frac{\cos\phi\, dS}{4\pi r^2}\right]$$

Integrating over all limits gives for the number of molecules hitting a unit area in unit time

$$\frac{dn}{dt} = n^* \int_{c=0}^{\infty} f(c)dc \int_{r=0}^{c} dr \frac{1}{4\pi} \int_{\phi=0}^{\frac{\pi}{2}} \sin\phi \cos\phi\, d\phi \int_{\theta=0}^{2\pi} d\theta$$

$$= n^* \int_{c=0}^{\infty} cf(c)dc \cdot \frac{\sin^2\phi}{8\pi}\Bigg|_0^{\frac{\pi}{2}} \theta \Bigg|_0^{2\pi}$$

$$= \tfrac{1}{4} n^* \bar{c}$$

This equation is similar to the previous developed value for dn/dt and differs only in the coefficient.

20.3 MEAN FREE PATH AND COLLISION FREQUENCY

The mean free path is defined as the average distance a molecule travels between collisions. Consider the following figure:

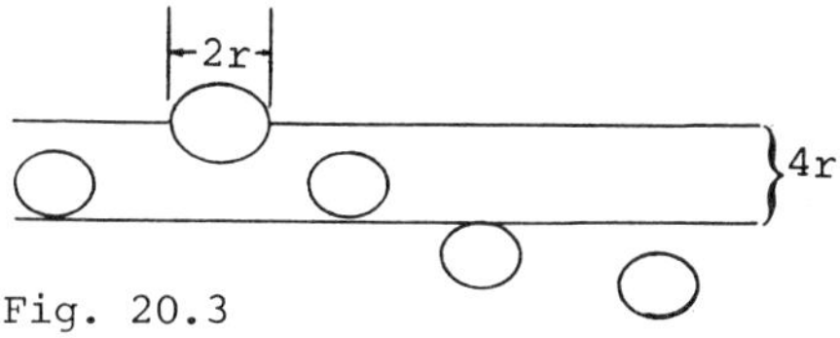

Fig. 20.3

a molecule with radius r sweeps out a cylindrical volume of diameter zr. As it moves, it will collide with other molecules, assuming that for the moment all the molecules are of the same diameter and that molecules other than the one in question are stationary (contained within a cylindrical volume with diameter 4r). That is, any molecule whose center of mass is less than 2r from that of the moving molecule will sustain a collision. The effective volume swept out in 1s is

$$4\pi r^2 \bar{c}$$

where $\bar{c}$ is the average speed and r is the molecular radius.

The molecule will collide with every other molecule in this volume. The number of such collisions is

$$Z_A = 4n^* \pi r^2 \bar{c} = n^* \pi d^2 \bar{c}; \quad d = 2r$$

where n* is again the number of molecules per unit volume.

The mean free path λ, is the distance traveled between collisions. This is simply the length of the path $\bar{c}$, divided by the number of collisions occurring in $\bar{c}$, or

$$\lambda = \frac{\bar{c}}{\pi n^* d^2 \bar{c}} = \frac{1}{\pi n^* d^2}$$

Of course the other molecules are not stationary and this must be considered. The $\bar{c}$ in the equation should actually be the average of the relative speeds of the colliding molecules.

$$\bar{C}_{AA} = \sqrt{2}\ \bar{C}$$

where $\bar{C}_{AA}$ is the average relative speed of Ⓐ,Ⓐ molecular collisions and $\bar{C}$ is simply the average speed of the Ⓐ molecules. Then

$$Z_A = \sqrt{2}\ n^* \pi\ d^2 \bar{c}$$

$$\lambda = \frac{1}{\sqrt{2}\pi\ n^* d^2}$$

and the total number of collisions between all molecules per unit volume is

$$Z_{total} = \frac{\sqrt{2}\,\pi\,\bar{c}\,d^2 n^{*2}}{2}$$

per second.

The factor of $\frac{1}{2}$ removes the twofold degeneracy in the number of actual collisions (i.e. A and A' collisions are the same as A' and A collisions).

A more general treatment is to consider a gas mixtue of N_1 molecules with molecular mass m_1 and N_2 molecules with molecular mass m_2. The internuclear distance is the sum of the individual radii, $r_1 + r_2 = d_{1,2}$, and the average relative speed between the two types of molecules is $\bar{C}_{1,2}$. The number of collisions between molecules ① and ② per unit volume is then

$$Z_{1,2} = \frac{\pi d_{1,2}^2 \bar{C}_{1,2} N_2}{V}$$

and the total collision frequency between molecules 1 and 2 is

$$Z_{1,2} = \frac{\pi\, d_{1,2}^2\ \bar{C}_{1,2}\, N_1 N_2}{V^2}$$

The number of ① molecules with velocity components between u_1 and $u_1 + du_1$, v_1 and $v_1 + dv_1$, w_1 and $w_1 + dw_1$ is

$$dN_1 = N_1 \left(\frac{m_1}{2\pi kT}\right)^{\frac{3}{2}} e^{-\left[\frac{m_1(u_1^2+v_1^2+w_1^2)}{2kT}\right)} du_1 dv_1 dw_1$$

The number of ② molecules with velocity components between u_2 and $u_2 + du_2$, v_2 and $v_2 + dv_2$, w_2 and $w_2 + dw_2$ is

$$dN_2 = N_2 \left(\frac{m_2}{2\pi kT}\right)^{\frac{3}{2}} e^{-\left[\frac{m_2(u_2^2+v_2^2+w_2^2)}{2kT}\right]} du_2 dv_2 dw_2$$

The number of collisions in unit volume in unit time between molecules ① and ② with velocity components in the respective ranges is then

$$Z_{1,2} = \frac{dN_1 dN_2 \pi d_{1,2}^2 C_{1,2}}{V^2} = \frac{N_1 N_2 \pi d_{1,2}^2 \bar{C}_{1,2}}{V^2}$$

$C_{1,2}$ is the relative speed of the molecules 1 and 2 in the specified velocity ranges.

$$C_{1,2} = [(u_1 - u_2)^2 + (v_1 - v_2)^2 + (w_1 - w_2)^2]^{\frac{1}{2}}$$

To calculate the total number of collisions in unit time between the ① and ② molecules, we substitute the expressions for dN_1, dN_2, C_{12} into the equation for Z_{12} and integrate over all values of $u_1, v_1, w_1, u_2, v_2, w_2$.

$$Z_{1,2} = \frac{1}{8} N_1 N_2 \pi d_{1,2}^2 \frac{(m_1 m_2)^{\frac{3}{2}}}{(\pi kT)^3} \iiiint\!\!\iint_{-\infty}^{\infty} [(u_1-u_2)^2 +$$

$$(v_1-v_2)^2 + (w_1-w_2)^2]^{\frac{1}{2}}$$

$$\cdot\, e^{\frac{-[m_1(u_1^2+v_1^2+w_1^2)+m_2(u_2^2+v_2^2+w_2^2)]}{2kT}}$$

$$\cdot\, du_1 dv_1 dw_1 du_2 dv_2 dw_2$$

To perform the above integration, the variables in $u_1, v_1, w_1, u_2, v_2, w_2$ must be transformed into center of mass coordinates. This procedure is easily demonstrated using the simple one-dimensional case.

$$E_{kinetic} = \tfrac{1}{2}m_1 v_1^2 + \tfrac{1}{2}m_2 v_2{}^2$$
$$= \tfrac{1}{2}(m_1 + m_2)V^2 + \tfrac{1}{2}\mu v_{1,2}^2$$

where

$$V_{1,2} = V_1 - V_2$$

$$\mu = \frac{m_1 m_2}{m_1 + m_2}$$

$$V = \frac{m_1 v_1 + m_2 v_2}{m_1 + m_2}$$

Then in terms of the center of body coordinates and the components of relative speed.

$$U = \frac{m_1 u_1 + m_2 u_2}{m_1 + m_2}; \quad V = \frac{m_1 v_1 + m_2 v_2}{m_1 + m_2}; \quad W = \frac{m_1 w_1 + m_2 w_2}{m_1 + m_2}$$

$$u_{1,2} = u_1 - u_2; \quad v_{1,2} = v_1 - v_2; \quad w_{1,2} = w_1 - w_2$$

$$e^{-[m_1(u_1{}^2+v_1{}^2+w_1{}^2)+m_2(u_2{}^2+v_2{}^2+w_2{}^2)]}$$

$$= e^{-[(m_1+m_2)(u^2+v^2+w^2)+\mu(u_{1,2}^2+v_{1,2}^2+w_{1,2}^2)]}$$

and the integrals become

$$Z_{1,2} = \frac{1}{8} N_1 N_2 \pi d_{1,2}^2 \frac{(m_1 m_2)^{\frac{3}{2}}}{(\pi kT)^3} \iiint_{-\infty}^{\infty} e^{\frac{-[(m_1+m_2)(u^2+v^2+w^2)]}{2kT}} dudvdw$$

$$\cdot \iiint_{-\infty}^{\infty} (u_{1,2}^2 + v_{1,2}^2 + w_{1,2}^2)^{\frac{1}{2}} e^{-[\mu(u_{1,2}^2+v_{1,2}^2+w_{1,2}^2)]} du_{1,2}dv_{1,2}dw_{1,2}$$

$$du_1 dv_1 dw_1 du_2 dv_2 dw_2 = dUdVdWdu_{1,2}dv_{1,2}dw_{1,2}$$

We prove this relation for one pair of components since the argument is applicable to the others as well. When changing from variables U_1, U_2 to the new variables $U_{1,2}(U_1,U_2)$ and $U(U_1,U_2)$, the products of the differentials are related by

$$du_{1,2}dU = \frac{\partial(u_{1,2},u)}{\partial(u_1,u_2)} du_1 du_2$$

where $\partial(u_{1,2},u)/\partial(u_1,u_2)$ is the Jacobian of the transform.

$$\frac{\partial(u_{1,2},u)}{\partial(u_1,u_2)} = \begin{vmatrix} \frac{\partial u_{1,2}}{\partial u_1} & \frac{\partial U}{\partial u_1} \\ \frac{\partial u_{1,2}}{\partial u_2} & \frac{\partial U}{\partial u_2} \end{vmatrix} = \begin{vmatrix} 1 & \frac{m_1}{m_1 + m_2} \\ -1 & \frac{m_2}{m_1 + m_2} \end{vmatrix} = 1$$

The first three integrals are of the form

$$\int_{-\infty}^{\infty} e^{-\frac{(m_1 + m_2)u^2}{2kT}} du = \left(\frac{2\pi kT}{m_1 + m_2}\right)^{\frac{1}{2}}$$

The product of the first three integrals is then

$$\iiint_{-\infty}^{\infty} e^{\frac{-[(m_1+m_2)u^2+v^2+w^2]}{2kT}} dudvdw = \left(\frac{2\pi kT}{m_1 + m_2}\right)^{\frac{3}{2}}$$

The second triple integral can be evaluated after transforming to spherical coordinates

$$U_{1,2}^2 + V_{1,2}^2 + W_{1,2}^2 = C_{1,2}^2$$

$$du_{1,2}dv_{1,2}dw_{1,2} = C_{1,2}^2 \sin\phi \, dc_{1,2} d\phi \, d\theta$$

so that the triple integral reduces to the more manageable form

$$\int_{C_{1,2}=0}^{\infty} \int_{\phi=0}^{\pi} \int_{\theta=0}^{2\pi} C_{1,2}^3 \, e^{\frac{-\mu c_{1,2}^2}{2kT}} \sin\phi \, d\phi d\theta \, dc_{1,2}$$

$$= 4\pi \int_{c_{1,2}=0}^{\infty} C_{1,2}^3 \, e^{\frac{-\mu c_{1,2}^2}{2kT}} dc_{1,2}$$

$$= 8\pi \left(\frac{kT}{\mu}\right)^2$$

Collecting results we finally have

$$Z_{1,2} = \frac{1}{8} N_1 N_2 \pi d_{1,2}^2 \frac{(m_1 m_2)^{\frac{3}{2}}}{(\pi kT)^3} \left(\frac{2\pi kT}{m_1 + m_2}\right)^{\frac{3}{2}} 8\pi \left(\frac{kT}{\mu}\right)^2$$

$$Z_{1,2} = N_1 N_2 d_{1,2}^2 \left(\frac{8kT}{\pi\mu}\right)^{\frac{1}{2}} = \pi d_{1,2}^2 N_1 N_2 \bar{c}_{1,2}$$

For the special case in which all the molecules are the same, $m_1 = m_2 = m$, $N_1 = N_2 = \frac{N}{2}$, $\mu = m/2$, $d_{1,2} = d = 2r$

$$Z_{1,1} = \frac{N^2}{2} \pi d^2 \sqrt{2} \left(\frac{8kT}{\pi m}\right)^{\frac{1}{2}} = \frac{\sqrt{2}}{2} \pi d^2 N^2 \bar{C}$$

which agrees with the previous development.

20.4 VISCOSITY

Viscosity is a measure of the frictional resistance that a fluid in motion offers to an applied shearing force. In the case of streamline flow, a fluid is divided into successive laminae for the purpose of analysis.

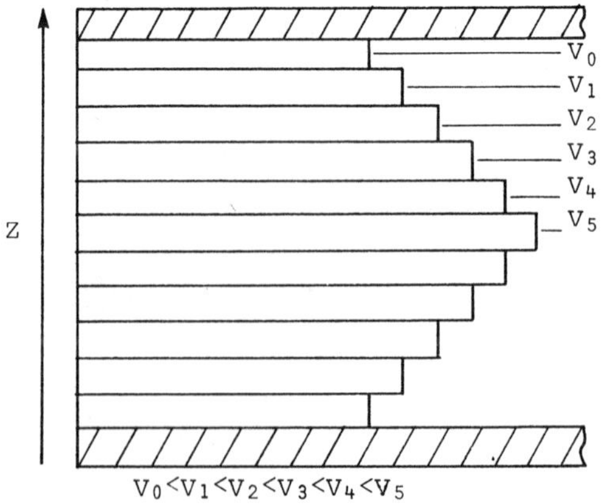

Fig. 20.4 (A)

If a fluid flows past a stationary plane surface, the lamina immediately adjacent to the surface is considered stagnant, while successive laminae have increasingly higher velocities. In the region of Newtonian flow, (i.e. immediately above and below the stagnant boundaries) the velocity varies linearly from lamina to lamina. Observations show that the frictional force f, resisting the relative motion of two adjacent layers, is proportional to A, the area of the interfacial region, and to dv/dz, the velocity gradient

$$f = -\eta A \frac{dv}{dz}$$

where η is called the coefficient of viscosity and its value is characteristic of the fluid. The negative sign indicates that the force is opposed to the motion.

The kinetic picture of gas viscosity can be represented by the following analogy: two trains are moving in the same direction on parallel tracks but at different speeds. Passengers amuse themselves by jumping back and forth from one train to the other. A passenger from the more rapidly moving train transports momentum $m\Delta V$, where m is his mass and ΔV is the excess velocity. He tends to speed up the more slowly moving train when he lands on it. Likewise, the passenger from the slower moving train tends to slow the faster train down when landing on it. The net effect is to equalize the speeds of the two trains. An outside observer who was not aware of the jumpers might simply note this as a frictional drag between the trains. The mechanism by which one layer or lamina of a gas exerts a viscous drag on an adjacent layer is similar in that the gas moelcules play the role of the errant passengers.

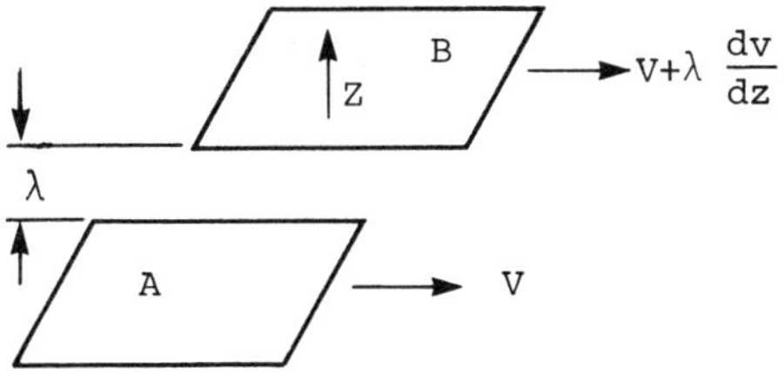

Fig. 20.4 (B)

Consider two adjacent lamina separated by a distance of one mean free path. In the region of Newtonian flow, the laminae velocities are linearly dependent and a molecule passing from A to B will on average bring momentum $-m\lambda \frac{dv}{dz}$ to B while a molecule traveling from B to A brings momentum

$m \lambda \frac{dv}{dz}$. As momentum is transferred from one layer to the next the directed motion of flow, v, is degraded into random thermal motion and as a result the temperature of the gas will rise. The number of molecules crossing unit area from above and below per unit time is twice $\frac{1}{4}n^*\bar{c}$ or $\frac{1}{2}n^*\bar{c}$. The momentum transport per unit time is then $\frac{1}{2}n^*\bar{c}m\lambda\frac{dv}{dz}$. This momentum change per unit time is equivalent to the frictional force per unit area.

$$f = -\eta A \frac{dv}{dz}$$

$$F = -\eta \frac{dv}{dz} = \frac{1}{2} n^*\bar{c}m\lambda \frac{dv}{dz}$$

$$\eta = \frac{1}{2} n^*\bar{c}\, m\lambda$$

$$\lambda = \frac{1}{\sqrt{2}\, n^* d^2 \pi}$$

$$\eta = \frac{m\bar{c}}{2(2)^{\frac{1}{2}}\pi\, d^2}$$

As can be seen from the final relation, viscosity is independent of density and pressure and linearly dependent, $\bar{c} \alpha\ T^{\frac{1}{2}}$, on temperature.

The usual method for determining viscosity is to use the Poiseville equation:

$$\frac{dv}{dt} = \frac{\pi (P_1^2 - P_2{}^2)R^4}{16l\eta P_0}$$

where P_1 is the high pressure, P_2 is the low pressure, R and l the dimensions of the capillary which are the radius and the length respectively, and P_0 is the pressure at which the volume is measured. dv/dt is the volume rate of flow.

The units of η in the sI system are $KGm^{-1}s^{-1}$. Vis-cosities are often expressed in centipoise.

$$1cp = 10^{-3}\ KGm^{-1}s^{-1}$$

20.5 THERMAL CONDUCTIVITY

Just as viscosity depends on the transport of momentum across a velocity gradient, thermal conductivity of a gas is a consequence of the transport of kinetic energy across a temperature gradient. The heat flow per unit time is related to the temperature gradient and a unit cross sectional area by

$$\frac{q}{t} = \kappa A \frac{dT}{dz}$$

where κ is the thermal conductivity coefficient.

Again we divide the gas into a series of layers, assuming the separation between layers to be equal to the mean free path. The average temperature difference between adjacent layers is $\lambda(dT/dx)$. If the molecules have mass m and a specific heat capacity C_v (heat capacity per unit mass), then the energy difference between molecules in two adjacent layers is $mc_v \lambda \left(\frac{dT}{dz}\right)$. The number of molecules passing through unit area in unit time is $\frac{1}{2}n^*\bar{c}$ (accounting for those traveling both up and down), so that the total energy transferred per unit area per unit time becomes

$$\frac{q}{At} = \frac{1}{2}n^*\bar{c}\ mc_v\ \lambda \left(\frac{dT}{dz}\right)$$

From this it follows

$$\kappa = \frac{1}{2}n^*\bar{c}mc_v\ \lambda = \eta c_v$$

The sI units of κ are $JK^{-1}m^{-1}s^{-1}$.

20.6 DIFFUSION

It is known experimentally that the flux of particles is proportional to the gradient of their concentration

$$J = -D\frac{dc}{dx}$$

where J is the flux per unit area per unit time, c is the concentration as a function of position, and D is the diffusion coefficient.

The preceding equation is known as Fick's first law of diffusion. In the process known as self diffusion, the mean free path treatment yields for the diffusion coefficient

$$D = -\tfrac{1}{2}\lambda\,\bar{c} = \frac{\eta}{\rho}$$

This is seen by allowing for the total mass transfer across a unit area.

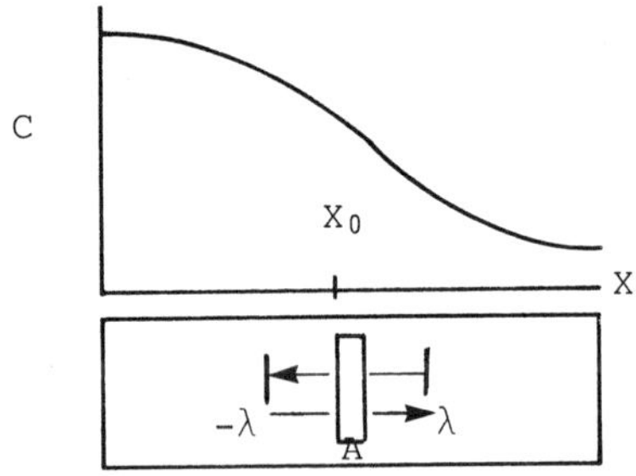

Fig. 20.5

The concentration of mass at $-\lambda$ is $C(x_0) - \lambda\left(\frac{dc}{dx}\right)_{x_0}$ and at λ is $C(x_0) + \lambda\left(\frac{dc}{dx}\right)_{x_0}$. The average number of collisions occurring on a unit area is $\frac{1}{4}\bar{c}\,\lambda(C(x_0) - \lambda\left(\frac{dc}{dx}\right)_{x_0})$ and $\frac{1}{4}\bar{c}\,\lambda(C(x_0) + \lambda\left(\frac{dc}{dx}\right)_{x_0})$ from each respective region. Since the flow from the more concentrated region dominates the backwash, the total flux J_x becomes

$$J_x = \tfrac{1}{4}\bar{c}\lambda\{(C(x_0) - \lambda\left(\frac{dc}{dx}\right)_{x_0}) - (C(x_0) + \lambda\left(\frac{dc}{dx}\right)_{x_0})\}$$

$$J_x = -\tfrac{1}{2}\bar{c}\lambda\left(\frac{dc}{dx}\right)_0$$

from which by comparison to Fick's first law the diffusion

coefficient can be deduced. For the case of a mixture of two different gases the diffusion coefficient becomes

$$D = \tfrac{1}{2}\lambda_1\bar{c}_1x_1 + \tfrac{1}{2}\lambda_2\bar{c}_2x_2$$

where λ_1 and λ_2 are the mean free paths of the respective gases, $\bar{c}_1$ and $\bar{c}_2$ are the respective average speeds, and x_1 and x_2 are the mole fractions.

CHAPTER 21

STATISTICAL MECHANICS

21.1 ENTROPY AND DISORDER

The entropy of a system is a measure of its tendency to become disordered. Statistically speaking, for a given collection of molecules there are many more unique distinguishable random distributions than ordered ones. For this reason, the most probable state of a system tends to be the more disordered one. To see how entropy is related to the probability of a particular state, consider the expansion of 1 mole of gas from a volume V_1 at pressure P_1 into an evacuated volume V_2. The final volume is $V_1 + V_2$ and the final pressure is $P_{external}$. The change in entropy is given as

$$\Delta S = S_2 - S_1 = R \ln \left(\frac{V_1 + V_2}{V_1}\right) = k \ln \left(\frac{V_1}{V_1 + V_2}\right)^{-L}$$

where R = gas constant and

k = Boltzmann constant.

The probability of finding all 'L' molecules in V_1 is:

$$P(V_1) = \left(\frac{V_1}{V_1 + V_2}\right)^L$$

The probability of finding all 'L' molecules in $(V_1 + V_2)$ is:

$$P(V_2) = 1^L = 1$$

The expression for the entropy change in terms of probabilities then becomes

$$\Delta S = k \ln \frac{P(V_2)}{P(V_1)} = k \ln \frac{\Omega_{final}}{\Omega_{initial}}$$

The relation between the entropy of a state and its probability is thus seen to be a logarithmic one. More generally it is expressed

$$S = k \ln \Omega + b; \quad b = \phi$$

where k is the Boltzmann constant, Ω the probability of the state, and from third law considerations the constant of integration, b, is taken to be zero.

21.2 LAGRANGE'S METHOD FOR CONSTRAINED EXTREMA AND THE STIRLING APPROXIMATION

We must introduce a mathematical method known as Lagrange's method of undetermined multipliers which will prove useful in later discussions. Given some function of n independent variables, $f(x_1, x_2, \ldots, x_n)$, subject to some additional constraints on those variables, such as $g(x_1, x_2, \ldots, x_n) = \phi$, we proceed as follows. In the absence of any constraint we would have for the total derivative of the original function becomes

$$\delta f = \sum_{i=1}^{n} \frac{\partial f}{\partial x_i} \delta x_i = \phi$$

for an extrema. The condition

$$g(x_1, x_2, \ldots, x_n) = \phi$$

implies that the n variables are no longer independent. Lagrange's method reduces the problem to one with n - 1 independent variables by introducing an undetermined multiplier. Given the variation in the constraint

$$\delta g = \sum_{i=1}^{n} \frac{\partial g}{\partial x_i} \delta x_i = \phi$$

Then by Lagrange's method

$$\sum_{i=1}^{n} \left(\frac{\partial f}{\partial x_i} + \lambda \frac{\partial g}{\partial x_i} \right) \delta x_i = \phi$$

where $$\frac{\partial f}{\partial x_\alpha} + \lambda \frac{\partial g}{\partial x_\alpha} = \phi$$

for an arbitrarily chosen dependent variable x_α. The multiplier is then

$$\lambda = \frac{-\partial f/\partial x_\alpha}{\partial g/\partial x_\alpha}$$

and the n - 1 coefficients of the remaining independent variables vanish separately

$$\frac{\partial f}{\partial x_i} + \lambda \frac{\partial g}{\partial x_i} = \phi; \quad i \neq \alpha$$

so that with λ given as above

$$\frac{\partial f}{\partial x_i} + \lambda \frac{\partial g}{\partial x_i} = \phi$$

is now valid for all x_i. The method is easily extended to situations with more than one constraint and generally the number of constraints determines the number of multipliers needed.

In computing the number of permutations of different possible arrangements of a large set of independent elements, it is often necessary to evaluate the factorials of rather large numbers. A useful formula for accomplishing this task is the Stirling approximation

$$\ln N! \underset{N \to \infty}{\approx} N \ln N - N$$

21.3 ENSEMBLES

An ensemble is a mental construct made for the purpose of mathematical analysis, of the statistical mechanical properties of a system. The two types most commonly encountered are the cannonical ensemble and the microcannonical ensemble.

The cannonical ensemble consists of a large number of systems each having the same value of N, V and T, where

N - is the number of particles making up the basic system

V - is the volume of the basic system

T - is the temperature.

and separated so that energy is allowed to pass from system to system without mass flow. At equilibrium each member of the ensemble will have the same temperature but not necessarily the same energy, E_i. The value of E_i will vary from member to member but will fluctuate around average. The total energy of the ensemble is constant, which allows a great number of possible distributions of its members.

The microcannonical ensemble is a similar construction in that N, V and now E, the energy of each system comprising the ensemble is the same. With a little thought it can easily be concluded that a cannonical ensemble is one member of a microcannonical ensemble. In the limit as N, the number of members of the microcannonical ensemble, approaches infinity, every possible state of the cannonical ensemble having an equal A priori probability is represented. Therefore, the average of any mechanical variable (i.e. P,T,S) of the original system over a long period of time, its equilibrium value, can be set equal to the ensemble average. This is known as the ergodic hypothesis.

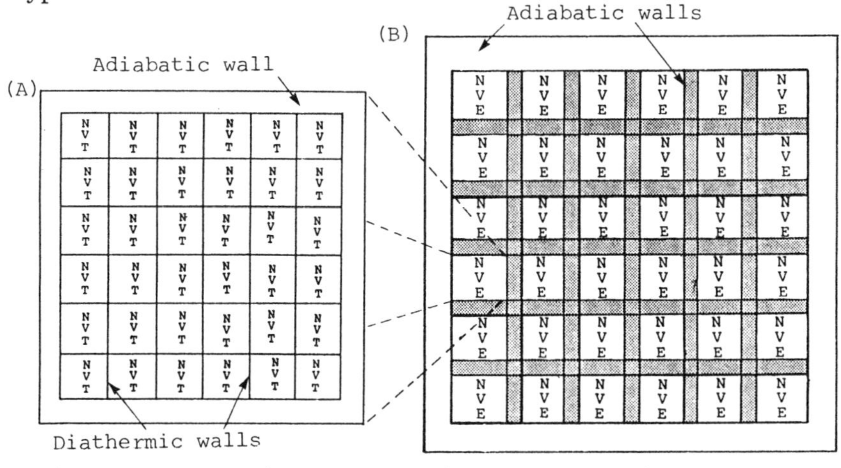

Fig 21.1 Schematic representation of a cannonical ensemble as one member of a micro cannonical ensemble.

21.4 THE BOLTZMANN DISTRIBUTION

As an example of using ensembles to calculate statistical properties, let us employ the technique to calculate the Boltzmann distribution.

A litre of an ideal gas can be considered to be a cannonical ensemble where V is one litre, T is specified by thermal equilibrium between other molecules, and N is 1. Each molecule has the same set of allowed energy states available, ε_1, ε_2, ε_3,..., etc. Since N, V, and E, the total energy of the litre of gas, are fixed, the litre of gas is one member of a microcannonical ensemble consisting of N members. With every A priori state accounted for, we can now proceed to calculate ensemble averages and set them equal to time average values. There are two constraints on each member of the microcannonical ensemble:

$$\Sigma N_j \varepsilon_j = E_{total} \qquad (1)$$

$$\Sigma N_j = N \qquad (2)$$

where N_j are the number of molecules with energy ε_j, and ΣN_j is one litre of molecules $\left(\approx \frac{6.02 \times 10^{23}}{22.4} \right)$.

One of a number of ways of arranging the molecules in 1 litre subject to the above constraints is

$$\Omega_n = \frac{N!}{N_1!N_2!\ldots N_j!} = \frac{N!}{\prod_j N_j!}$$

Ω_n is the probability of this particular state dependent upon the distributions set, $\prod_j N_j!$, which itself is dependent upon the constraints of mass and energy conservation. The total number of distinguishable states is the sum of W_n over all possible distribution sets.

$$W = \Sigma W_n = \sum_N \frac{N!}{\prod_j N_j!}$$

Consider as an example, two possible distrbutions of four molecules over four energy levels subject to the constraints

$$\sum N_j \varepsilon_j = 7\varepsilon$$

$$\sum N_j = 4$$

and where $\varepsilon_1 = \varepsilon$, $\varepsilon_2 = 2\varepsilon$, $\varepsilon_3 = 3\varepsilon$, etc.

Table 21.1

ε_1	ε_2	ε_3	ε_4		ε_1	ε_2	ε_3	ε_4
abc			d		ab	c	d	
abd			c		ab	d	c	
acd			b		ac	b	d	
bcd			a		ac	d	b	
					ad	b	c	
					ad	c	b	
					bc	a	d	
					bc	d	a	
					bd	a	c	
					bd	c	a	
					cd	a	b	
					cd	b	a	

$$\frac{4!}{3!0!0!1!} = 4 \qquad \frac{4!}{2!1!1!0!} = 12$$

Permutations of particles in similar energy states are irrelevent

$$\underset{ab}{\varepsilon_1} = \underset{ba}{\varepsilon_1}$$

For an ensemble with a large number of particles, the average values of the set of N_j can be set equal to the most probable values of N_j, for the system. The most probable set of N_j will be those that maximize Ω_n with respect to all possible variations of the N_j's. Since the natural log of a function is monotonic (single-valued), we can maximize the log of this function and achieve the same result.

$$\Omega_n = \frac{N!}{\prod_j N_j!}$$

$$\ln \Omega_n = \ln N! - \sum_j \ln N_j!$$

$$\delta \ln \Omega_n = 0 = \sum_j \delta \ln N_j! \quad N \text{ is constant } \therefore \delta(\ln N!) = 0$$

From Stirling's approximation

$$\delta \sum_j \ln N_j! = \delta \sum_j N_j \ln N_j - \delta \sum_j N_j = 0$$

Now using the method of Lagrange on the constraints

$$\delta N = \sum_j \delta N_j = 0$$

$$\delta E = \sum_j \varepsilon_j \delta N_j = 0$$

we obtain

$$\sum_j \ln N_j \delta N_j + \sum_j \delta N_j - \sum_j \delta N_j + \alpha \sum_j \delta N_j + \beta \sum_j \varepsilon_j JN_j = 0$$

so that

$$\ln N_j + \alpha + \beta \varepsilon_j = 0 = \frac{\delta \Omega_n}{\delta N_j}$$

for the independent variables and finally

$$N_j = e^{-\alpha - \beta \varepsilon_j}$$

since

$$\sum_j N_j = N$$

We may eliminate the constant α

$$\frac{N_j}{N} = \frac{e^{-\alpha - \beta\varepsilon_j}}{e^{-\alpha} \sum_j e^{-\beta\varepsilon_j}} = \frac{e^{-\beta\varepsilon_j}}{\sum_j e^{-\beta\varepsilon_j}}$$

This is the Boltzmann distribution.

The constant β can be computed by using the fundamental equation of thermodynamics.

$$dS = \frac{dE}{T} + \frac{PdV}{T} - \frac{\mu}{T} dN$$

where

$$\left(\frac{\partial S}{\partial E}\right)_{V,N} = \frac{1}{T}$$

Evaluating $\left(\frac{\partial S}{\partial E}\right)_{V,N}$ from the relation

$$S = k \ln \Omega$$

and setting the two equal gives

$$\Omega = \frac{N!}{\prod_j N_j!}$$

$$S = k(\ln N! - \sum_j \ln N_j!)$$

$$= k(N \ln N - N - \sum_j N_j + \sum_j N_j)$$

with $\sum_j N_j = N \sum_j \varepsilon_j N_j = E$

$$\ln N_j = -(\alpha + \beta\varepsilon_j)$$

$$S = k(N \ln N + \alpha N + \beta E)$$

and $\alpha = -\ln N + \ln \sum_j e^{-\beta\varepsilon_j}$

so that $S = k(N \ln \sum_j e^{-\beta\varepsilon_j} + \beta E)$

Now taking $\left(\frac{\partial S}{\partial E}\right)_{V,N}$

$$\left(\frac{\partial S}{\partial E}\right)_{V,N} = k\left(-N \frac{\sum_j \varepsilon_j e^{-\beta\varepsilon_j}}{\sum_j e^{-\beta\varepsilon_j}} \frac{d\beta}{dE} + E\frac{d\beta}{dE} + \beta\right)$$

$$= k\left(-N \frac{\sum_j \varepsilon_j e^{-\beta\varepsilon_j}}{\sum_j e^{-\beta\varepsilon_j}} + E\right)\frac{d\beta}{dE} + k\beta$$

The first term on the right is zero. This can be shown by multiplying

$$\frac{\sum_j \varepsilon_j e^{-\beta\varepsilon_j}}{\sum_j e^{-\beta\varepsilon_j}}$$

by $e^{-\alpha}$ in the numerator and denominator.

$$\frac{\sum_j \varepsilon_j e^{-\alpha-\beta\varepsilon_j}}{\sum_j e^{-\alpha-\beta\varepsilon_j}} = \frac{\sum_j \varepsilon_j N_j}{\sum_j N_j} = \frac{E}{N}$$

Substituting this back into the original equation leaves

$$\left(\frac{\partial S}{\partial E}\right)_{V,N} = -k\beta$$

but from the fundamental equation of thermodynamics

$$\left(\frac{\partial S}{\partial E}\right)_{V,N} = \frac{1}{T} \quad \therefore \quad \beta = \frac{1}{kT}$$

The Boltzmann distribution becomes

$$\frac{N_j}{N} = \frac{e^{-\varepsilon_j/kT}}{\sum_j e^{-\varepsilon_j/kT}}$$

The denominator is called the molecular partition function and is given the symbol Z.

$$Z = \sum_j e^{-\varepsilon_j/kT}$$

In the event there is a degeneracy, that is several states having the same energy, a statistical weighting factor must be included in the Boltzmann distribution.

$$\frac{N_j}{N} = \frac{g_j e^{-\varepsilon_j/kT}}{\sum_j g_j e^{-\varepsilon_j/kT}}$$

Assume for the moment that the spacing between energy levels is of the order kT. The distribution for a system of one mole of particles is then

Table 21.2

Number of Level (u)	$\frac{\varepsilon_j}{kT}$	$e^{-\varepsilon_j/kT}$	Number of molecules in each level $N = 6.02 \times 10^{23}$ $g_j = 1$	$g_j = j$
1	0	1	3.806×10^{23}	2.407×10^{23}
2	1	0.3679	1.400×10^{23}	1.771×10^{23}
3	2	0.1353	0.515×10^{23}	0.977×10^{23}
4	3	0.0498	0.190×10^{23}	0.479×10^{23}
5	4	0.0183	0.070×10^{23}	0.220×10^{23}
6	5	0.0067	0.030×10^{23}	0.097×10^{23}
7	6	0.0025	0.0095×10^{23}	0.042×10^{23}
8	7	0.0009	0.0034×10^{23}	0.017×10^{23}
9	8	0.0003	0.0011×10^{23}	0.006×10^{23}
10	9	0.000123	0.0005×10^{23}	0.003×10^{23}
$Z = \sum_j g_j e^{-\varepsilon_j/kT}$			1.58182	2.50123

Observe that for the non-degenerate case roughly 60% of the molecules are in the lowest energy state.

21.5 EQUATIONS OF STATE AND THE MOLECULAR PARTITION FUNCTION

The entropy can be formally related to the partition function as

$$S = Nk \ln Z + \frac{E}{T}$$

from the previous development. It is, however, convenient to use the Helmholtz equation of state to develop equations of state in terms of the partition function

$$A = E - TS$$

with S given as above

$$A = -NkT \ln Z$$

For the ensemble average of energy

$$E = A + TS$$

$$= A - T\left(\frac{\partial A}{\partial T}\right)_{V,N}$$

$$= -T^2 \frac{\partial}{\partial T}\left(\frac{A}{T}\right)$$

where $$S = -\left(\frac{\partial A}{\partial T}\right)_{V,N}$$

Making the appropriate substitutions gives

$$E = NkT^2 \frac{\partial}{\partial T} (\ln Z)_{V,N}$$

Additional thermodynamic functions can be obtained as follows

$$P = -\left(\frac{\partial A}{\partial V}\right)_{V,N} = NkT\left(\frac{\partial \ln Z}{\partial V}\right)_{T,N}$$

$$C_V = \left(\frac{\partial E}{\partial T}\right)_{V,N} = \frac{\partial}{\partial T}\left(NkT^2 \frac{\partial \ln Z}{\partial T}\right)_{V,N}$$

$$= \frac{Nk}{T^2}\left(\frac{\partial^2 \ln Z}{\partial \left(\frac{1}{T}\right)^2}\right)_{V,N}$$

21.6 EVALUATING THE MOLAR PARTITION FUNCTION

Given the previous equations and the partition function

Z, it would seem that the values of any equilibrium property of matter could be calculated. Z, however, can only be evaluated for systems with non-interacting particles (i.e. an ideal gas), if serious mathematical difficulties are to be avoided. If we consider the total energy of a system -1 mole of a monatomic gas for instance - to be the sum of the energies of its individual constituents we would have

$$E_{total} = \sum_{i,j=1}^{L,M} \varepsilon_{i,j} = \varepsilon_{1,j} + \varepsilon_{2,j} + \varepsilon_{3,j} + \ldots + \varepsilon_{L,j}$$

where i is the subscript denoting a particular molecule and is summed over all L molecules (6.02×10^{23}), and j is the subscript denoting the particular energy level, from the M available levels, that the ith molecule occupies.

Recalling the definition of the molecular partition function

$$z = \sum_{j} e^{-\varepsilon_j/kT}$$

and assuming each molecular constituent is distinguishable from all the remaining ones gives

$$Z = \prod_{i}^{L} \sum_{j}^{M} e^{-\varepsilon_{i,j}/kT}$$

$$= (\sum_{j} e^{-\varepsilon_{1,j}/kT})(\sum_{j} e^{-\varepsilon_{2,j}/kT})\ldots(\sum_{j} e^{-\varepsilon_{L,j}/kT})$$

$$= z_1 z_2 \ldots z_L$$

The product of the individual molecular Z's generates all possible states of the 'L' molecules. 'Z' is known as the molar partition function. Since the same set of energy levels is available for each of the 'L' molecules

$$Z = z^L$$

we must correct this expression so as not to count certian states too many times. Terms of the type

$$e^{-(\varepsilon_{1,i} + \varepsilon_{2,j} + \varepsilon_{3,k} + \ldots)/kT}$$

where $i \neq j \neq k$, will occur L! times in the summation because the L molecules can be permuted L! times over the L states (for simplicity, it is assumed that the number of available levels just equals the number of molecules).

$$Z = \frac{z^L}{L!}$$

21.7 THE TRANSLATIONAL PARTITION FUNCTION

To calculate Z, we need to know the allowed energy states of the molecule. These energy states can be determined experimentally from detailed spectroscopic data. Suffice it to say

$$Z = \sum_{n_1=0}^{\infty} \sum_{n_2=0}^{\infty} \sum_{n_3=0}^{\infty} e^{-[\frac{h^2}{8mkT}(\frac{n_1^2}{a^2} + \frac{n_2^2}{b^2} + \frac{n_3^2}{c^2})]}$$

where n_1, n_2, n_3 are summed over all states, h is Planck's constant, and a, b and c are the dimensions of the space within which a particle of mass m is confined. (Bear in mind we are dealing with, translational kinetic energy.)

Since h is so small, the summation can be replaced by integrals

$$Z = \int_0^{\infty}\int_0^{\infty}\int_0^{\infty} e^{-[\frac{h^2}{8mkT}(\frac{n_1^2}{a^2} + \frac{n_2^2}{b^2} + \frac{n_3^2}{c^2})]}$$

which with

$$I = \int_0^{\infty} e^{(-A^2x^2)}dx = \frac{\sqrt{\pi}}{2A}$$

reduccs the triple integral to a product of three integrals with

$$A^2 = \frac{h^2}{8ma^2kT}$$

Similar integrals are found for the b and c terms.

$$Z = \frac{(2\pi mkT)^{\frac{3}{2}}}{h^3}(abc) = \frac{(2\pi mkT)^{\frac{3}{2}}}{h^3}V$$

The molar partition function is then

$$Z = \frac{Z^L}{L!} = \frac{1}{L!}\left[\frac{(2\pi mkT)^{\frac{3}{2}}V}{h^3}\right]^L$$

and we now check to see if the statistical formulations agree with macroscopic results.

$$A = -kT\ln Z = kT\ln L! - \frac{3}{2}kLT\ln(2\pi mkT) - kLT\ln V + 3kLT\ln h$$

$$P = -\left(\frac{\partial A}{\partial V}\right)_T = \frac{kLT}{V} = \frac{RT}{V}$$

which is the ideal gas law for 1 mole.

$$E = kT^2\left(\frac{\partial \ln Z}{\partial T}\right)_{V,N} = LkT^2\{\frac{3}{2}\frac{1}{T}\} = \frac{3}{2}RT$$

which is the classical result from the principle of equipartition of energy.

$$S = -\left(\frac{\partial A}{\partial T}\right)_{V,N} = kT\left(\frac{\partial \ln Z}{\partial T}\right)_{V,N} + k\ln Z$$

$$S = RT\{\frac{3}{2}\frac{1}{T}\} + k\ln\frac{1}{L!}\left[\frac{(2\pi mkT)^{\frac{3}{2}}V}{h^3}\right]$$

From Stirling's approximation

$$\ln N! \approx N\ln N - N \quad \therefore \quad L! = \left(\frac{L}{e}\right)^L$$

$$S = \frac{3}{2}R + k\ln\left[\frac{(2\pi mkT)^{\frac{3}{2}}Ve}{h^3 \quad L}\right]^L = R\ln\left[\frac{(2\pi mkT)^{\frac{3}{2}}Ve^{\frac{5}{2}}}{h^3L}\right]$$

where the $\frac{3}{2}$ term has been conveniently included in the log expression.

This equation is known as the Sakur and Tetrode equation. To see that it does in fact agree with classical values, we use it to calculate the change in entropy for an isothermal expansion of an ideal gas.

$$\Delta S = S_2 - S_1 = R\left\{ \ln\left[\left(\frac{2\pi mkT}{h^2}\right)^{\frac{3}{2}} \frac{V_2 e^{\frac{5}{2}}}{L}\right] - \ln\left[\left(\frac{2\pi mkT}{h^2}\right)^{\frac{3}{2}} \frac{V_1 e^{\frac{5}{2}}}{L}\right]\right\} = R\ln\frac{V_2}{V_1}$$

CHAPTER 22

MATTER AND WAVES

22.1 SIMPLE HARMONIC MOTION

The simplest model of a harmonic oscillator is a mass, m, attached to a rigid support by a spring with a force constant K. The spring is assumed to be perfectly elastic - i.e. no viscous damping forces - and to have no mass itself. When displaced from its equilibrium position, the mass experiences a restoring force opposite in direction to the displacement force and proportional to the distance from the equilibrium position. Using Newton's Law

$$F = ma$$

and equating forces gives the following equation. Here -Kx is the restoring force.

$$\boxed{ma = m\frac{d^2x}{dt^2} = -Kx}$$

Here K = force constant

x = displacement of spring.

This is a simple linear differential equation. It can be solved by making the substitutions

$$v = \frac{dx}{dt};\quad \frac{d^2x}{dt^2} = \frac{dv}{dt} = \frac{dv}{dx}\,\frac{dx}{dt}$$

Then $\frac{d\ x}{dt} = V\frac{dv}{dx}$

$$v \frac{dv}{dx} + \frac{k}{m} x = 0$$

so that $$v^2 + \frac{k}{m} x^2 = C$$

At x = A, i.e. at amplitude of vibration, the velocity is zero and c is

$$c = \frac{k}{m} A^2$$

$$\left(\frac{dx}{dt}\right)^2 = v^2 = \frac{k}{m} (A^2 - x^2)$$

$$\int \frac{dx}{[(A^2 - x^2)]^{\frac{1}{2}}} = \int \left(\frac{k}{m}\right)^{\frac{1}{2}} dt + c'$$

From a standard integral table, the integral on the left can be found

$$\sin^{-} \frac{x}{A} = \left(\frac{k}{m}\right)^{\frac{1}{2}} t + c'$$

At x = 0, t = 0 so that c' = 0. Finally

$$x = A\sin \left(\frac{k}{m}\right)^{\frac{1}{2}} t$$

which is the solution of the equation of motion for a harmonic oscillator. If we set

$$\left(\frac{k}{m}\right)^{\frac{1}{2}} = 2\pi\nu$$

where ν is the frequency of the motion we get

$$x = A\sin 2\pi\nu t$$

The simple harmonic motion (S.H.M.) can be represented graphically by a sine curve.

The quantity 'A' is the amplitude of the vibration. At 'A' since V = 0, the kinetic energy is zero, and all the energy is potential. At x = 0, all the energy is kinetic. The total energy is the sum of the kinetic and potential energies

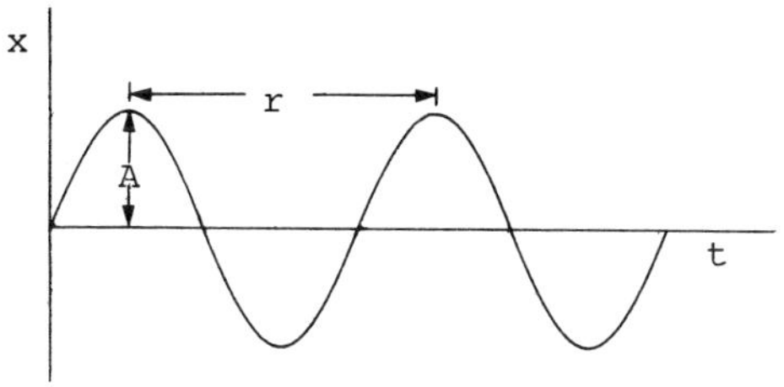

Simple harmonic vibration, displacement x as a function of time t.

Fig. 22.1

$$E = E_p + E_k$$

Since the total energy is constant, it must equal the potential energy at 'A'. The potential energy of the oscillator is given as

$$E = \frac{1}{2}KA^2$$

The total energy is proportional to the square of the amplitude. This relation holds true for all periodic motions.

22.2 WAVE MOTION

The previously described motion is an oscillatory motion but not a wave motion since there is no propagation of energy. A simple example of wave motion is the displacement of a transverse wave in a string. The displacement of the string at any instant can be represented by

$$u = f(x,t)$$

If we focus attention on the displacement at one point, x, and if the propagation of this displacement has velocity v in the positive x direction, then at a time t

$$u = f(x,t) = f(x - vt)$$

Suppose the wave at t = 0 has the profile

$$u = A\sin 2\pi \frac{x}{\lambda}$$

where A is the amplitude, and λ is the wavelength.

The moving sine wave then becomes

$$u = A\sin \frac{2\pi}{\lambda} (x - vt)$$

since $\nu = \frac{v}{\lambda}$. This can also be written as, where ν is now the frequency,

$$u = A\sin 2\pi \left(\frac{x}{\lambda} - \nu t \right)$$

If we differentiate 'u' twice with respect to t and twice with respect to 'x' we find that

$$\frac{\partial^2 u}{\partial x^2} = \frac{1}{v^2} \frac{\partial^2 u}{\partial t^2}$$

This is the general partial differential equation for wave motion in one dimension.

22.3 STANDING WAVES

Suppose the wave motion previously described is propagated over a length of string, L. We would then have what is known as a boundary value problem. In many cases in mathematical physics the solution of a differential equation will contain some parameter, the values of which must be selected in order to meet the boundary conditions. These selected values are called Eigenvalues. The solutions that correspond to the Eigenvalues are called Eigenfunctions.

The solution to the wave equation, can be expressed as the product of the functions of the independent variables, since it is a linear differential equation,

$$U(x,t) = X(x)T(t)$$

Then

$$\frac{\partial^2 u}{\partial x^2} = T(t)\,\frac{\partial^2 X}{\partial x^2}$$

$$\frac{\partial^2 u}{\partial t^2} = X(x)\,\frac{\partial^2 T}{\partial t^2}$$

so that

$$\frac{1}{X}\,\frac{\partial^2 X}{\partial x^2} = \frac{1}{v^2 T}\,\frac{\partial^2 T}{\partial t^2}$$

The only way the left side of this equation will always equal the right side is if both sides are equal to the same constant which we shall call $\frac{-\omega^2}{v^2}$,

$$\frac{\partial^2 T}{\partial t^2} = -\omega^2 T; \quad \frac{\partial^2 X}{\partial x^2} = \frac{-\omega^2}{v^2}\,X$$

the solutions of which are

$$T = e^{\pm i\omega t} \qquad X = e^{\pm i\omega x/v}$$

Thus

$$U = X(x)T(t) = e^{\pm i\omega x/v}e^{\pm i\omega t}$$

This equation can be rewritten in the form of real sine and cosine functions

$$U = \begin{matrix}\sin\\ \cos\end{matrix}\,\omega\left(t \pm \frac{x}{v}\right) \quad \text{or} \quad \begin{matrix}\sin\\ \cos\end{matrix}\,(\omega t)\begin{matrix}\sin\\ \cos\end{matrix}\left(\frac{\omega x}{v}\right)$$

For $u = 0$ at $x = 0$, the cosine function of x must be deleted, giving

$$U = \sin\left(\frac{\omega x}{v}\right)\begin{matrix}\sin\\ \cos\end{matrix}(\omega t)$$

In order that $U = 0$ at $x = L$

$$\sin\frac{\omega L}{v} = 0 \quad \text{for} \quad \frac{\omega L}{v} = n\pi$$

where n is an integer. Then for the solution of the wave equation we get

$$u_n = \begin{matrix}\sin\\ \cos\end{matrix}\,\omega_n t\,\sin\frac{n\pi\alpha}{L}$$

Either the sine or the cosine function, or any desired combination, can be used as the function of t.

$$u_n = (A_n \sin \omega_n t + B_n \cos \omega_n t) \sin \frac{n\pi x}{L}$$

The solution above represents what is called a standing wave. The function of x always has the same form irrespective of the value of t. Strictly speaking, a standing wave is an oscillatory motion where there is an interchange between kinetic and potential energy similar to the S.H.O. All the standing waves must satisfy the condition

$$n \frac{\lambda}{2} = L$$

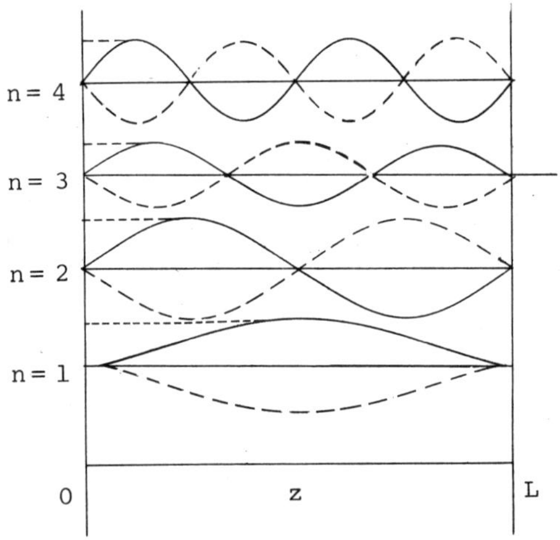

Fig. 22.3 Blackbody radiation and the nuclear atom.

This is the simplest form of the eigenvalue condition. The functions u_n are then the eigenfunctions for the problem.

22.4 BLACKBODY RADIATION AND THE NUCLEAR ATOM

One of the most significant developments in our understanding of the structure of the atom was provided by Ernest Rutherford in 1911. Experimental investigations of α particle scattering in thin metal foils seemed to indicate that atoms were composed of extremely dense positive nuclei which contained nearly all of an atom's mass. The electrons necessary to maintain electrostatic neutrality would then have to be distributed in the remaining volume of the atom with orbits determined by the balancing of both the electrical attractive forces directed inward and the centrifugal forces directed outward. However, according to classical electromagnetic theory, electrons revolving about the nucleus are accelerated charged particles and should continuously emit radiation, lose energy, and spiral downward into the nucleus.

In addition to the inability of classical electromagnetic theory to explain the seemingly paradoxical nature of atomic structure, the wave theory of light was being questioned for its inability to adequately explain the phenomena of black body radiation.

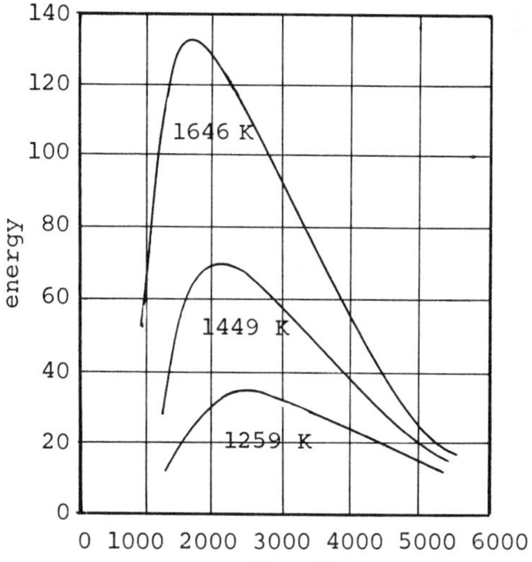

Fig. 22.4 (A)

Experimental measurements of Lummer and Pringsheim on spectral distribution of radiation from a black body at three different temperatures.

All objects are continually absorbing and emitting radiation. For a body to be in equilibrium with its environment, the radiation emitted must be equivalent (in wavelength and energy) to the radiation it is absorbing.

It is possible to conceive of objects that are perfect absorbers of radiation, the so-called ideal black bodies. The best laboratory black body is an insulated cavity in which a small orifice is made in one of the walls. When the cavity is heated the orifice emits radiation of continuous wavelengths. At any given temperature there is a characteristic distribution of frequencies.

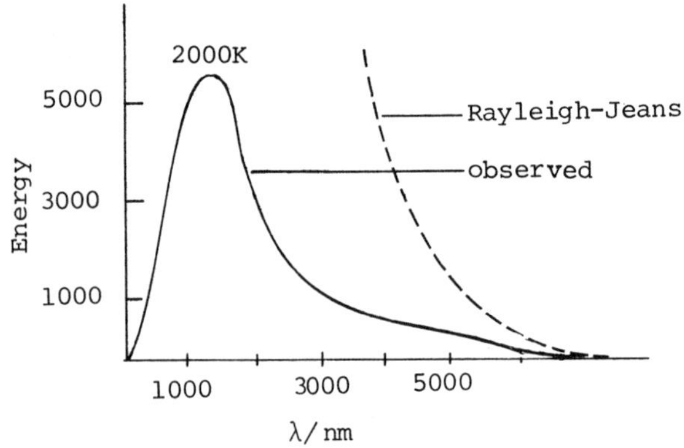

Fig. 22.4 (B)

Attempts to explain the observed distributions based on the principle of the equipartition of energy were unsuccessful and led to what was then called the ultraviolet catastrophe. According to the equipartition principle, an oscillator in thermal equilibrium should have an average energy of kT, $\frac{1}{2}kT$ for kinetic energy and $\frac{1}{2}kT$ for potential energy. The average energy should in no way depend on the frequency of the oscillator. In a system containing 100 oscillators, 20 with a frequency ν_1 and 80 with a frequency ν_2, where $\nu_2 > \nu_1$, the equipartition principle predicts that 20% of the total energy should be in the low frequency oscillators and 80% in the high frequency oscillators. Therefore the intensity of radiation in the range of wavelength λ to $\lambda + d\lambda$ is the number of oscilators in that range multiplied by their average energy. An explicit calculation gives the Rayleigh-Jeans Law

$$dU(\lambda) = \frac{8\pi kT}{\lambda^4} d\lambda$$

which suggests that as λ decreases, - x-rays, γ-rays - oscillators within the range λ and $\lambda + d\lambda$ are strongly excited at room temperature as absurd as the results are, their conclusions are unavoidable when classical mechanics is used.

22.5 THE PHOTOELECTRIC EFFECT

In 1887, H. Hertz observed that a spark would jump a gap more readily when the gap electrodes were illuminated by light from another spark gap. The phenomena known as the photoelectric effect was adequately explained by Albert Einstein in 1905. Based on the assumption that radiation is quantized, he concluded that the kinetic energy of an ejected electron was dependent on the frequency of incident radiation

$$\tfrac{1}{2}mv^2 = h\nu - e\phi$$

where h is Planck's constant and ϕ is the work function of the material in question.

Three observations which led to this conclusion were

A) No electrons were ejected, regardless of the intensity of the incident radiation, unless a characteristic threshold frequency was exceeded.

B) The kinetic energy of the ejected electrons was proportional to the frequency of the incident radiation.

C) Even at low intensities, electrons were ejected immediately if the threshold frequency was slightly exceeded.

These observations, especially the last one, indicated that the energy of the radiation was not distributed classically over a broad wave front but localized in discrete packets subsequently named photons.

22.6 SPECTROSCOPY AND THE BOHR ATOM

Throughout the nineteenth century it was believed

that line spectra of atoms were produced simultaneously by individual atoms behaving as an oscillator with a large number of different periods of vibration. The atomic hydrogen line in the visible region of the spectrum, however, seemed to folow the regular relationship

$$\frac{1}{\lambda} = R\left(\frac{1}{z^2} - \frac{1}{n_1^2}\right)$$

where R is the Rydberg constant = $109{,}677.581 cm^{-1}$ and $n_1 = 3,4,5$, etc.

Other hydrogen series obeyed the more general formula

$$\frac{1}{\lambda} = R\left(\frac{1}{n_2^2} - \frac{1}{n_1^2}\right)$$

where $n_2 = 1$ for the Lyman series, $n_2 = 3$ for the Paschen series, $h_2 = 4$ for the Brackett series, and $n_2 = 5$ for the Pfund series.

The evidence of the day seemed to indicate that V. Rutherford type atoms matter was composed of which when excited underwent quantized transitions of state with consequent absorption of energy according to the Planck-Einstein relation

$$\boxed{\varepsilon = h\nu = E_2 - E_1}$$

Bohr theorized that for an electron to remain in orbit around the nucleus

$$\frac{Ze^2}{4\pi\varepsilon_0 r^2} = \frac{mv^2}{r}$$

where e is the charge on the electron and ε_0 is the permitivity constant which led to the determination of the Bohr radius

$$r = \frac{Ze^2}{4\pi\varepsilon_0 m_e v^2}$$

The angular momentum of the electron was taken to be

$$mvr = \frac{nh}{2\pi}$$

where n is an integer which gives

$$r = \frac{\varepsilon_0 n^2 h^2}{\pi m_e e^2 Z}$$

where ε_0 is the permitivity of free space. A constant n is the principle quantum number; m_e and e are the mass and charge of the electron respectively and z is the atomic number.

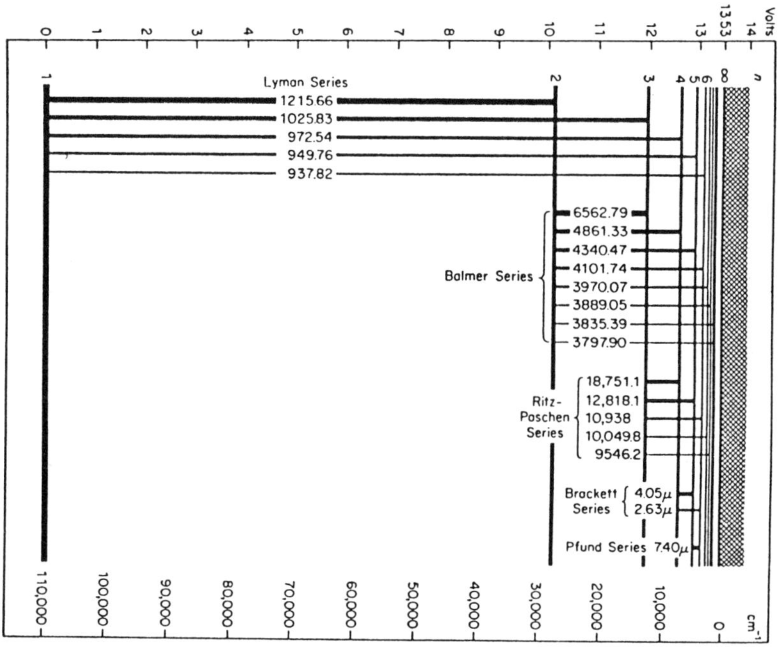

Fig. 22.6

In the case of hydrogen Z = 1, and the smallest orbit would be for n = 1

$$a_0 = \frac{\varepsilon_0 h^2}{\pi m_e e^2} = .05292\text{nm}$$

The energy levels are calculated as follows

$$E = E_{kinetic} + E_{potential} = \frac{mv^2}{2} - \frac{Ze^2}{4\pi\varepsilon_0 r}$$

$$E = \frac{-Ze^2}{4\pi\varepsilon_0 r} + \frac{Ze^2}{8\pi\varepsilon_0 r} = \frac{-Ze^2}{8\pi\varepsilon_0 r}$$

Substituting the value for r gives

$$E = \frac{-m_e e^4}{8\varepsilon_0^2 h^2} \frac{Z^2}{n^2}$$

The frequency of a spectral line due to the transition between quantum levels N_1 and N_2 is

$$\nu = \left(\frac{1}{h}\right)(E_{n_1} - E_{n_2}) = \frac{m_e e^4 Z^2}{8\varepsilon_0^2 h^3} \cdot \left(\frac{1}{n_2^2} - \frac{1}{n_1^2}\right)$$

This expression yields a theoretical value for the Rydberg constant, since

$$\boxed{\nu = \frac{c}{\lambda}}$$

$$R = \frac{m_e e^4}{8\varepsilon_0^2 c h^3} = 109{,}737 \text{cm}^{-1}$$

m in the preceding equation should be corrected to represent the reduced mass of the system μ.

$$\mu = \frac{m_e m_p}{m_e + m_p}$$

where m_e and m_p are the masses of the electron and the proton respectively.

The energy levels become more closely packed in the higher quantum states until they converge into a region of continuous absorption without spectral line structure. Once the electron is completely free from the attractive forces of the nucleus, it is no longer restricted to quantized energy states and may continuously take up kinetic energy of translation. The difference in energy between the series limit and the ground state is called the ionization potential and if atoms contain more than one electron, there will be first, second, third, etc. ionization potentials.

22.7 THE DE BROGLIE RELATION

The overwhelming conclusion that the energy in atomic

systems was restricted to integral values suggested a link to another branch of physics in which integral numbers occur naturally, namely the stationary state solutions of the equation for wave motion. This would imply that electrons, and in fact all matter, possess wavelike properties.

In 1923 de Broglie showed this to be true theoretically. The necessary condition for a stable orbit of radius r_e for an electron orbiting a nucleus is

$$\boxed{2\pi r_e = n\lambda}$$

where n is an integer.

This condition arises from the assumption that only integral numbers of wavelengths will form standing wave patterns without incurring destructive interference.

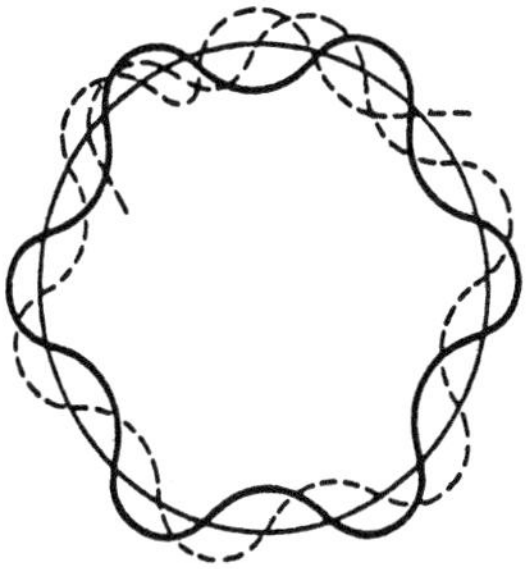

Fig. 22.7

The solid line represents a possible stationary wave. The dashed line shows a wave of different wavelength that would be destroyed by interference.

In the case of the photon, two fundamental equations are to be obeyed:

$$\varepsilon = h\nu; \qquad \varepsilon = mc^2$$

Equating the two terms leads to

$$h\nu = mc^2$$

and with

$$\nu = \frac{c}{\lambda}$$

$$\lambda = \frac{h}{mc}$$

Now for electron waves, de Broglie considered a similar equation governed wavelength. Thus

$$\lambda_e = \frac{h}{m_e v_e} = \frac{h}{P_e}$$

where v_e is the velocity of the electron and P_e is the momentum.

Substituting into the original standing wave equation the assumed value of λ gives

$$m_e v_e r_e = \frac{nh}{2\pi}$$

which is the original Bohr condition for a stable orbit. The de Broglie relation

$$\lambda = \frac{h}{r}$$

is the fundamental relation between the momentum of a particle and its wavelength.

CHAPTER 23

QUANTUM MECHANICS

23.1 THE SCHRÖDINGER EQUATION

The mathematical development of quantum chemistry begins with the Schrödinger's equation. Strictly speaking, the wave equation cannot be derived from any more fundamental postulates. It occupies in quantum chemistry a position analogous to that of Newton's F = ma in classical mechanics. The general differential equation of wave motion was previously given as

$$\boxed{\frac{\partial^2 u}{\partial x^2} = \frac{1}{v^2}\frac{\partial^2 u}{\partial t^2}}$$

where $U = u(x,t)$

The solutions were shown to be variable separable and to take the form

$$u(x,t) = w(x)e^{2\pi irt}$$

where $\omega = 2\pi r$ in the T(t) function.

Direct substitution of this function into the wave equation gives

$$e^{2\pi irt}\left(\frac{d^2 w}{\partial x^2} + \frac{4\pi^2 v^2}{v^2}w\right) = 0$$

or

$$\frac{d^2 w}{dx^2} + \frac{4\pi^2 v^2}{v^2}w = 0$$

This is the wave equation with the time dependence removed. To apply this equation we use the de Broglie relation as follows: The total energy of a particle with mass m is the sum of its potential, and kinetic energies.

$$E = \frac{P^2}{2m} + v \quad \text{where } P = mv$$

also

$$P = [2m(E - v)]^{\frac{1}{2}}$$

Now using the de Broglie relation

$$\lambda = \frac{h}{P} = \frac{h}{[2m(E - v)^{\frac{1}{2}}}$$

since

$$v = \frac{V}{\lambda}$$

$$v^2 = \frac{2mv^2(E - v)}{h^2}$$

and now we substitute this value into the wave equation giving

$$\frac{d^2\psi}{x^2} + \frac{8\pi^2 m}{h^2}(E - v)\psi = 0$$

where w has been replaced by ψ.

This is the Schrödinger's equation in one dimension. For three dimensions it becomes

$$\overline{\nabla}^2\psi + \frac{8\pi^2 m}{h^2}(E - v)\psi = 0$$

where

$$\overline{\nabla}^2 = \frac{\partial^2}{\partial x^2} + \frac{\partial^2}{\partial y^2} + \frac{\partial^2}{\partial z^2}$$

The Hamiltonian operator is defined as

$$H = \frac{-h^2}{8\pi^2 m}\overline{\nabla}^2 + v$$

In terms of the Hamiltonian, the Schrödinger equation can be written simply as

$$H\psi = E\psi$$

The solutions must satisfy particular boundary conditions imposed upon the system.

Just as the standing wave equation yields discrete solutions when the eigenvalue condition is satisfied, so appropriate solutions are obtained for the Schrödinger equation for specific energy values. Discrete energy values are formed whenever a particle is constrained to move in a deferred space, wereas a continuous range of values for E is found for the electron moving freely through space. The allowed energies are called the eigenvalues for the system and the corresponding wave functions are called the eigenfunctions.

If $\psi(x)$ is a solution of the wave equation, then the relative probability of finding an electron within the range x to x + dx is given as

$$\psi(x) = \psi^*(x)dx,$$

where ψ^* is the complex conjugate of the wave function if it has complex components otherwise it is identically equal to ψ, and is found by replacing i with -i, or -i with i respectively. In order for the wave function to be interpreted as a probability amplitude it must obey certain mathematical conditions.

23.2 POSTULATES OF QUANTUM MECHANICS

The postulates upon which a logical development of quantum mechanics are based for the single particle system without spin properties are

A) The physical state of a particle as a function of time is described as fully as possible by a wave function

$$\psi(x,t)$$

B) The wave function and its first and second derivatives $\frac{\partial \psi}{\partial x}$, $\frac{\partial^2 \psi}{\partial x^2}$, must be continuous, finite, and single-valued for all values of x.

C) Any quantity that is physically observable can be represented in quantum mechanics by a Hermitian operator. Hermitian operators are linear operators that satisfy the condition for any pair

$$\int \psi_1^* \hat{H} \psi_2 dx = \int \psi_2 (\hat{H}\psi_1)^* dx$$

of functions, ψ_1, ψ_2, which represent the physical states of the particle (note the asterisk (*) denotes the complex conjugate of a function, i.e. $Z = x + iy$ and $Z^* = x - iy$).

D) The physically observable quantities of a system are given by the eigenvalues of the Hermitian operator

$$\hat{O}\psi_i = O_i \psi_i$$

E) The expectation value of an observable $<O>$ corresponding to the operator $\hat{O}$, is calculated from the formula

$$\int_{-\infty}^{\infty} \psi^* \hat{O} \psi \, dx = <O> \equiv \underline{O}$$

which assumes the wave function to be normalized

$$\int_{-\infty}^{\infty} \psi^* \psi dx = 1$$

F) Table 23.1

Classical Variable	Quantum mechanics operator	Expression	Operation
x	$\hat{x}$	x	Multiply by x
P_x	$\hat{P}_x$	$\frac{\hbar}{i}\frac{\partial}{\partial x}$	take derivative and multiply by $\hbar/i$
t	$\hat{t}$	t	multiply by t
E	$\hat{E}$	$\frac{-\hbar}{i}\frac{\partial}{\partial t}$	take derivative and multiply by $-\hbar/i$

The quantum mechanical operator corresponding to a physical quantity is constructed by expressing the classical operator in terms of x, P_x, t and E, and converting that expression to an operator by means of the above rules.

G) The wave function $\psi(x,t)$ is a solution of the time dependent Schrödinger equation

$$\hat{H}(x,t)\ \psi(x,t) = \frac{i\hbar\partial\psi(x,t)}{\partial t}$$

where $\hat{H}$ is the Hamiltonian operator found by expressing the classical Hamiltonian by means of the correspondence rules in the preceding table.

23.3 OPERATORS

An operator is best described as an instruction to carry out a mathematical operation upon a function, which is called an operand. For example

$$f(x) = x^3$$

$$\hat{O} = \frac{\partial}{\partial x} \qquad \hat{O}f(x) \Rightarrow 3x^2$$

$$\hat{O} = \frac{\partial}{\partial x} \qquad \hat{O}_2 = x$$

$$\hat{O}_1\hat{O}_2 f(x) \Rightarrow 4x^3$$

The operators are then seen to be performed consecutively. If $\hat{O}_2\hat{O}_1 f(x) = \hat{O}_1\hat{O}_2 f(x)$, then the operators are said to commute. An operator is said to be linear for any two functions f and g if

$$\hat{O}(af + bg) = a(\hat{o}f) + b(\hat{o}g)$$

For example, $\frac{d^2}{dx^2}$ is a linear operator while $\sqrt{\ }$ is not.

If the expectation value $<\xi>$, of a linear operator

is to be real, (i.e. a phsycially observable value), it must equal its complex conjugate $<\xi>^*$. The complex conjugate is obtained by taking the complex conjugate of the wave function

$$<\xi>^* = \int_{-\infty}^{\infty} \psi\hat{\xi}^* \psi^* dx \quad \text{so that}$$

$$<\xi> = \int_{-\infty}^{\infty} \psi^* \hat{\xi}\psi \, dx = <\xi>^* = \int_{-\infty}^{\infty} \psi(\hat{\xi}\psi)^* dx$$

It is evident from this that the operator is Hermitian. For a function f and an operator $\hat{O}$

$$\hat{O}f = cf$$

where c is a number, then f is called an eigenfunction of the operator $\hat{O}$ and c is an eigenvalue of the operator. For Hermitian operators, the eigenvalues c must be real.

23.4 SOLUTIONS OF SCHRÖDINGER'S EQUATION

23.4.1 THE FREE PARTICLE

In the case of a free particle - i.e. one moving in the absence of any potential field - the one-dimensional wave equation becomes

$$\frac{d^2\psi}{dx^2} + \frac{8\pi^2 m_e}{h^2} E\psi = 0$$

where m_e is the mass of the electron and the equation represents electron wave motion the solution of which takes the form

$$\psi = A\exp(\pm 2\pi i \sqrt{2mE}\, x/h)$$

or, alternately in trigonometric form

$$\psi = A\sin(2\pi\sqrt{2mE}\ x/h)$$

$$\psi = A\cos(2\pi\sqrt{2mE}\ x/h)$$

the solutions may be verified by direct substitution for those unfamiliar with this type of differential equation.

From the observation that the sine and cosine functions have a period of 2π we see that

$$\frac{2\pi\sqrt{2mE}}{h} = \frac{2\pi}{\lambda}$$

and from the de Broglie relation

$$\lambda = \frac{h}{p} = \frac{h}{\sqrt{2mE}}$$

the momentum of the system then is

$$P_x = \pm\sqrt{2mE}$$

since the particle may either travel to the left or the right. There are no restrictions on E for the free particle and thus the kinetic energy in the unbound state can have any positive value.

23.4.2 THE PARTICLE IN A RING OF CONSTANT POTENTIAL

Consider a particle of mass "m" moving in a ring of radius, r. The particle has zero potential and requires infinite energy to get out of the orbit. In other words, the potential everywhere outside the orbit is infinite. The variable coordinate is the angle θ. Comparison of the formulas of linear momentum and angular momentum reveals that the analogous variables in circular motion are the moment of inertia and angular velocity.

$$I = mr^2$$

$$\omega = v/r,$$

note I is the moment of inertia and ω is the angular velocity

$$P_\theta = mvr = I\omega$$

$$P_x = mv \Rightarrow \hat{P}_\theta = \frac{-h^2}{8\pi^2 I}\frac{d^2}{d\theta^2}$$

the Schrödinger's equation for circular motion then becomes

$$\frac{-h^2}{8\pi^2 I}\frac{d^2\psi(\theta)}{d\theta^2} = E\psi(\theta)$$

the solution of which are

$$Ae^{\pm ik\theta} \quad \text{or} \quad A'\sin(k\theta) \quad \text{and}$$

$$A'\cos(k\theta) \quad \text{where } k = \frac{2\pi\sqrt{2IE}}{n}$$

(Students unfamiliar with second-order differential equations may verify the solution by substitution.) The energies of the particle are then

$$E = \frac{k^2h^2}{8\pi^2 I} \qquad k = 0,1,2,3,\ldots.$$

In order that the wavefront be single-valued, it is necessary that the function connect smoothly onto itself. This is equivalent to saying ψ and ψ' must repeat themselves everytime θ changes by 2π radians. Then with ψ given as $\sin(k\theta)$

$$\sin(k\theta) = s(k(\theta + 2\pi)) \text{ and for } \psi'$$

$$k\cos(k\theta) = k\cos(k(\theta + 2\pi))$$

These relations are satisfied if k is an integer. The same result follows if ψ is given in terms of the cosine function with the exception that the case k = 0 is not permitted since the function would vanish everywhere. k = 0 is, however, allowed for the cosine form. The normalized solutions are

$$\psi = \frac{1}{\sqrt{\pi}}\sin(k\theta) \qquad k = 1,2,3$$

$$\psi = \frac{1}{\sqrt{\pi}}\cos(k\theta) \qquad k = 1,2,3$$

$$\psi = \frac{1}{\sqrt{2\pi}} \qquad k = 0 \text{ for cosine case only}$$

The exponential form provides similar results

$$Ae^{\pm ik\theta} = Ae^{\pm ik(\theta + 2\pi)} = Ae^{\pm ik\theta}e^{\pm ik2\pi}$$

which gives $e^{\pm ik2\pi} = 1$

From Euler's relationship

$$e^{ix} = \cos x + i\sin x$$

and using the positive exponential set of solutions

$$e^{ik2\pi} = \cos(2\pi k) + i\sin(2\pi k) = 1$$

which implies k is again an integer. Thus, an alternate set of solutions for ψ is

$$\psi = \frac{1}{\sqrt{2\pi}} e^{\pm ik\theta} \qquad k = 0,1,2,3,....$$

We have then, a single state with E = 0 and doubly degenerate states above. The solution at E = 0 means that there is no finite zero point energy associated with free rotation. The double degeneracy of higher eigenfunctions reflects the fact that there are two possible orientations of motion on the ring.

23.4.3 THE PARTICLE IN A BOX

If we restrict the motion of the particle in one dimension to a well of infinite potential (i.e. the potential energy, v, at the boundaries o and L is ∞, while inside the well v = 0) the problem closely resembles the vibrating string fixed at both ends. To obtain standing waves, it was necessary to restrict the allowed wavelengths to half integral values between 0 and L.

$$n\frac{\lambda}{2} = L$$

In the same manner we proceed with the particle confined to a well of infinite potential.

$$\psi = A \sin kx + E \cos kx$$

$$\psi(0) = 0 \quad \text{and so} \quad B = 0$$

since sinx at x = 0 is 0.

$$\psi = A \sin kx, \quad k = \frac{2\pi}{\lambda} \quad \text{and with}$$

$$\psi = A \sin\left(\frac{2\pi\sqrt{2mE}}{n}\right)x, \qquad \lambda = \frac{h}{\sqrt{2mE}}$$

since v = 0. For $\psi(L) = 0$, we must have

$$\frac{2\pi}{n}(2mE)^{\frac{1}{2}}L = n\pi$$

where n is 1,2,3, etc. This condition restricts the allowed values of E to discrete eigenvalues.

$$E_n = n^2h^2/8mL^2$$

we must have

$$\frac{2\pi}{n}(2mE)^{\frac{1}{2}}L = n\pi$$

where n is 1,2,3, etc. This condition restricts the allowed values of E to discrete eigenvalues.

$$E_n = \frac{n^2h^2}{8mL^2}$$

Particle in a one-dimensional box. Allowed waves and energy levels.

It is apparent that as the value of L increases, the kinetic energy E_n decreases. The more room in which the electron has to move, the lower will be its kinetic energy. The more localized its motion is, the higher will be its kinetic energy. Since lower energy brings about greater stability in a system, the delocalization of the motion of a particle, such as an electron, in molecules with certain kinds of structures, notably conjugated and aromatic carbon compounds, leads to enhanced stability of the compound.

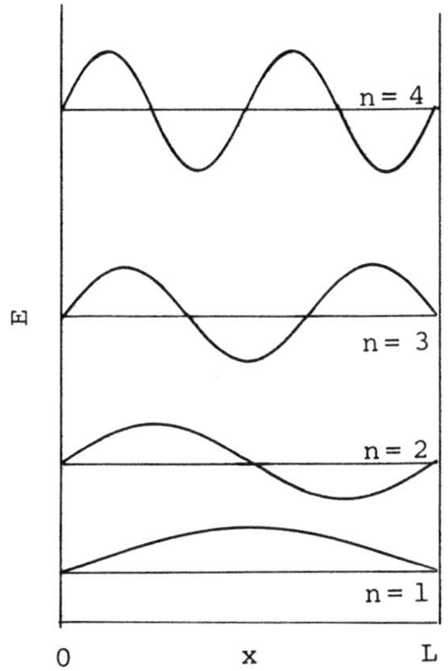

Fig. 23.1

The results of the solution of the Schrödinger's equation can easily be extended to three dimensions. Consider a parallelpiped with sides l, w, h. The potential v equals zero everywhere within the box. The Schrödinger's equation becomes for this system

$$H\psi \equiv \frac{-h^2}{8\pi^2 m}\left(\frac{\partial^2}{\partial x^2} + \frac{\partial^2}{\partial y^2} + \frac{\partial^2}{\partial z^2}\right)\psi = E\psi$$

the equation can be separated into

$$\psi = X(x)Y(y)Z(z),$$

which gives

$$\frac{-h^2}{8\pi^2 m}\left(\frac{1}{X}\frac{\partial^2 x}{\partial x^2} + \frac{1}{Y}\frac{\partial^2 Y}{\partial y^2} + \frac{1}{Z}\frac{\partial^2 Z}{\partial z^2}\right)$$

$$= E$$

Since this equation is true for all values of x,y,z, each term of the left-hand side of this equation must equal some constant.

$$\frac{-h^2}{8\pi^2 m}\left(\frac{1}{X}\frac{\partial^2 X}{\partial x^2}\right) = E_x$$

$$\frac{-h^2}{8\pi^2 m}\left(\frac{1}{Y}\frac{\partial^2 Y}{\partial y^2}\right) = E_y$$

$$\frac{-h^2}{8\pi^2 m}\left(\frac{1}{Z}\frac{\partial^2 Z}{\partial z^2}\right) = E_z$$

$$E = E_x + E_y + E_z$$

The separable equations are similar to the previously solved one-dimensional case, so that

$$X(x) = A_x \sin k_x x + B_x \cos k_x x$$

$$Y(y) = A_y \sin k_y y + B_y \cos k_y y$$

$$Z(z) = A_z \sin k_z z + B_z \cos k_z z$$

With the boundary conditions $\psi(x,y,z) = 0$ at $x = y = z = 0$, the cosine terms drop out if $B_x = B_y = B_z = 0$. From our original assumption of half integral wavelengths filling the box length

$$n\frac{\lambda}{2} = L$$

and the fact that $k = \frac{2\pi}{\lambda}$, we find for each length of the box $k = \frac{n\pi}{L}$, and subsequently

$$\psi_{n_l n_w n_h}(x,y,z) = A \sin\frac{n_l \pi x}{l} \sin\frac{n_w \pi y}{w} \sin\frac{n_h \pi z}{h}$$

where A is a combined constant of normalization involving l, w, and h.

$$E = E_x + E_y + E_z = \frac{h^2}{8m}\left(\frac{n_l^2}{l^2} + \frac{n_w^2}{w^2} + \frac{n_h^2}{h^2}\right)$$

(Note: h is Planck's constant, while h is a dimension of the box.)

The eigenvalues, E, for this three-dimensional problem depend on three distinct quantum numbers. To find A, we normalize the function

$$\int_0^l \int_0^w \int_0^h \psi^2(x,y,z)\,dx\,dy\,dz = 1$$

$$A^2 \int_{x=0}^{l} \int_{y=0}^{w} \int_{z=0}^{h} \sin^2 \frac{n_l \pi x}{l} \sin^2 \frac{n_w \pi y}{w} \sin^2 \frac{n_h \pi z}{h} dzdydx$$

$$= 1$$

$$\int_0^h \sin^2 \frac{n_h \pi z}{h} dz = \int_0^h \frac{1}{Z} \left(1 - \cos \frac{2n_h \pi z}{h} \right) dz$$

$$= \frac{Z}{2} \Big|_0^h - \frac{h}{4n_h \pi} \sin \frac{2n_h \pi z}{h} \Big|_0^h$$

$$= \frac{h}{2}$$

Integration of the terms remaining in y and x gives

$$A^2 \frac{hlw}{8} = 1$$

from which we find

$$A = \left(\frac{8}{V} \right)^{\frac{1}{2}}$$

where v = lxwxh.

If the box is cubical with l = h = w

$$E = \frac{h^2}{8ml^2} (n_1^2 + n_2^2 + n_3^2)$$

The occurrence of more than one distinct eigenfunction corresponding to the same eigenvalue for the energy is called degeneracy. For example, the eigenfunctions $\psi_{1,2,1}$, $\psi_{2,1,1}$, $\psi_{1,1,2}$ correspond to different distributions in space but all have the same energy

$$E = \frac{N^2 h^2}{8ml^2} \quad \text{where } N^2 = (n_1^2 + n_2^2 + n_3^2)$$

for the cubical box case. The energy level is said to be threefold degenerate.

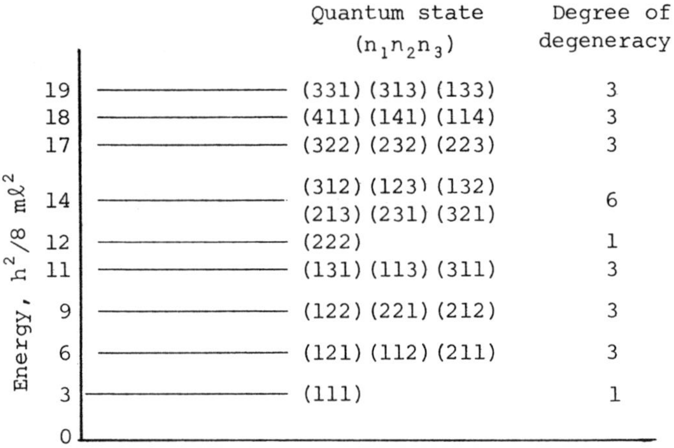

Fig. 23.2 Allowed energy levels for the particle in the cubic box.

23.4.4 THE ONE-DIMENSIONAL BOX WITH ONE FINITE WALL

Consider modifying the previous system by lowering the potential of one wall.

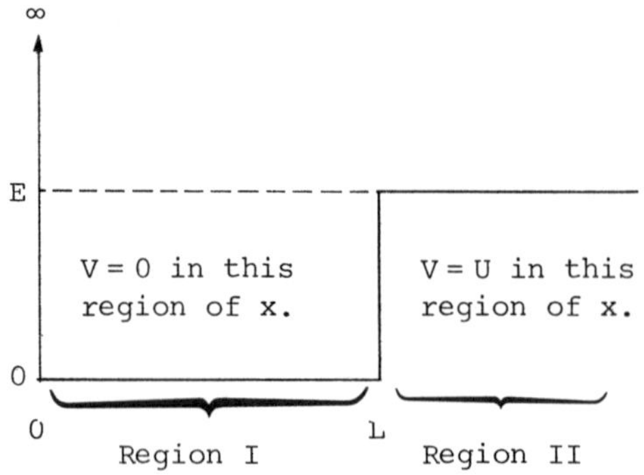

Fig. 23.3

When the particle is in Region I the general solution is

$$\psi_I = A_I \sin \frac{2\pi x}{\lambda_I} + B_I \cos \frac{2\pi x}{\lambda_F}$$

and B_I is zero from the boundary constraint $\psi_I(0) = 0$.

Therefore

$$\psi_I = A_I \sin \frac{2\pi x}{\lambda_I}$$

$$\lambda_I = \frac{h}{\sqrt{2mE}}$$

In region II there are two possibilities to consider $E > U$ or $E < U$. The general form of the solution is

$$\psi_{II} = C_{II} e^{\frac{+2\pi i x}{\lambda_{II}}} + D_{II} e^{\frac{-2\pi i x}{\lambda_{II}}}$$

which is just the variable form of the trigonometric solution. With $E < U$, λ_{II} if imaginary and becomes

$$\lambda_{II} = \frac{h}{\sqrt{2m(E - U)}}$$

and $(E - U)$ is negative. Assume λ_{II} is i times a positive number. Substituting this value of λ_{II} into the equation ψ_{II} we find that $e^{\frac{+2\pi i x}{\lambda_{II}}}$ is positive and expands without bound. It is therefore an inappropriate choice for a wave function and $C_{II} = 0$. The acceptable wavefunction becomes

$$\psi_{II} = D_{II} e^{\frac{-2\pi i x}{\lambda_{II}}}$$

Now, at $x = L$ the two wavefunctions of each region must join smoothly that is

$$\psi_I \Big|_L = \psi_{II} \Big|_L \quad \text{and} \quad \psi'_I \Big|_L = \psi'_{II} \Big|_L$$

$$A_I \sin \frac{2\pi L}{\lambda_I} = D_{II} e^{\frac{-2\pi i L}{\lambda_{II}}}$$

$$\frac{2\pi}{\lambda_F} A_I \cos\left(\frac{2\pi L}{\lambda_I}\right) = \left(\frac{-2\pi i}{\lambda_{II}}\right) D_{II} e^{\frac{-2\pi i L}{\lambda_{II}}}$$

from which we get

$$A_I \sin \frac{2\pi L}{\lambda_I} = \frac{-A_I \lambda_{II}}{i\lambda_I} \cos \frac{2\pi l}{\lambda_I}$$

Substituting the values of λ_I and λ_{II} and rearranging gives

$$\tan \frac{2\pi L\sqrt{2mE}}{h} = \frac{-\sqrt{E}}{\sqrt{U - E}}$$

For given values of L, m, and U, only certain values of $E < U$ will satisfy the equation above. Thus the particle can have only certain energies in this box. Soltuions for the values of E are found by graphing right and left hand members of the equation. the values of E where the plots intersect are the solutions.

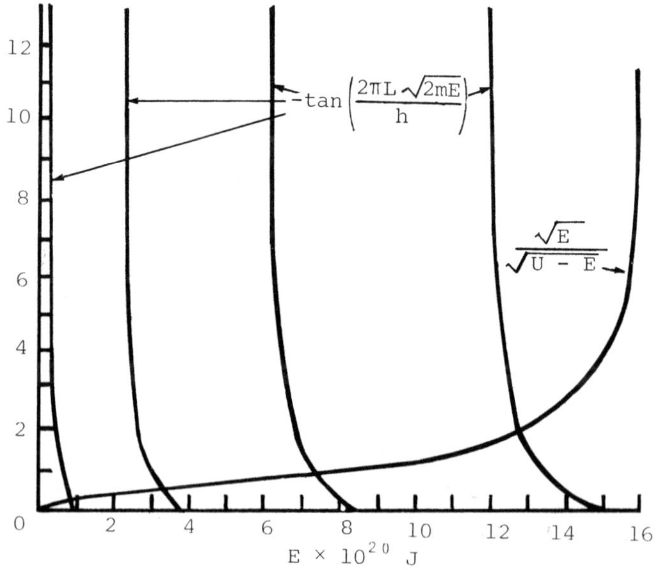

Fig. 23.4 Graphical solution of the equation $-\tan(2\pi L\sqrt{2mE}/h) = \sqrt{E}/\sqrt{U - E}$. Here L = 25 Å, $m = 9.11 \times 10^{-28}$ gm, $U = 1eV = 16.02 \times 10^{-20}$J. Intersections occur at $E = 0.82815 \times 10^{-20}$J, 3.29869×10^{-20}J, 7.35730×10^{-20}J and 12.82153×10^{-20}J.

A_I and D_{II} may be found by normalizing the respective wave functions. A set of solutions is shown in the following figure.

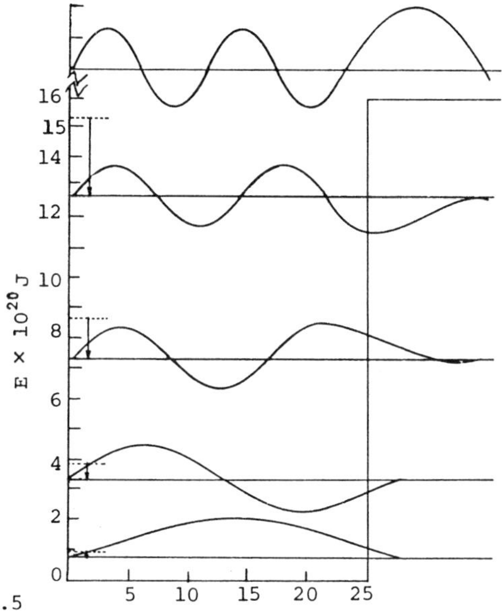

Fig. 23.5

As can be seen in the diagram, there is a finite probability that a particle in the well will penetrate the barrier. The ability of a particle to penetrate into potential regions where it classically was prohibited from entering is known as quantum mechanical tunneling.

In the case where E > U, ψ_I is a sine wave with wavelength determined by the de Broglie relation

$$\lambda_I = \frac{h}{\sqrt{2mE}}$$

where E is the kinetic energy of the particle and ψ_{II} has the solution

$$\psi_{II}(x) = A_{II}\sin\left(\frac{2\pi x}{\lambda_{II}}\right) + B_{II}\cos\left(\frac{2\pi x}{\lambda_{II}}\right)$$

where A_{II} and B_{II} can be adjusted to join ψ_{II} smoothly to ψ_I. ψ_{II} goes on oscillating as $x \to \infty$ with a wavelength slightly larger than ψ_I

$$\lambda_{II} = \frac{h}{\sqrt{2m(E - U)}} \qquad U \neq 0$$

CHAPTER 24

ROTATIONS AND VIBRATIONS OF ATOMS AND MOLECULES

24.1 HARMONIC OSCILLATOR

The one-dimensional harmonic oscillator model describes motion for a particle which alternately has its total energy stored in kinetic and potential energy. The potential energy function is given as

$$\boxed{V(x) = \tfrac{1}{2}Kx^2}$$

where $K = 4\pi^2 mV_0^2$

m is the mass of the system and v_0^2 is the fundamental vibration frequency.

K is analogous to a spring constant and is found from the solution of the differential equation of motion of the simple harmonic oscillator (Chapter 22) to be equal to the terms on the right-hand side of the preceding equation.

The Schrödinger equation for the system then becomes

$$\frac{-h^2}{8\pi^2 m}\frac{\partial^2\psi(x)}{\partial x^2} + \tfrac{1}{2}Kx^2\,\psi(x) = E\psi(x)$$

This equation can be simplified by substituting in the following relations:

$$\alpha^4 = \frac{\hbar^2}{Km}, \qquad \varepsilon = \frac{2\alpha^2 mE}{\hbar^2}$$

where
$$\hbar = \frac{h}{2\pi}$$

The Schrödinger's equation can then be rewritten

$$\alpha^2 \frac{\partial^2\psi}{\partial x^2} + \left(E - \frac{x^2}{\alpha^2}\right)\psi = 0$$

which can be verified by substitution. Next, transforming the independent variable x into a new variable y

$$X = \alpha y$$

and using the operator

$$\frac{\partial^2}{\partial x^2} = \frac{1}{\alpha^2}\frac{\partial^2}{\partial y^2}$$

gives
$$\frac{\partial^2\psi}{\partial y^2} + (E - y^2)\psi = 0$$

When y becomes very large, this reduces to

$$\frac{\partial^2\psi}{\partial y^2} = y^2\psi$$

and in the limit as $y \to \pm\infty$ the asymptotic solutions to this equation is

$$\psi = e^{\pm\frac{y^2}{2}}$$

The positive exponetial form is discarded since it does not behave properly, and a solution to the original equation takes the form

$$\psi = H(y)e^{-y^2/2}$$

When this expression is substituted into the Schrödinger's equation we obtain the differential equation that must be satisfied by H(y)

$$\frac{\partial^2 H}{\partial y^2} - 2y\frac{\partial H}{\partial y} + (\varepsilon - 1)H = 0$$

We can express H(y) as a power series in y around the

regular point $y = 0$.

$$H(y) = \sum_v a_v y^v \equiv a_0 + a_1 y + a_2 y^2 + \ldots.$$

with

$$\frac{\partial H}{\partial y} = \sum_v v a_v y^{v-1} \equiv a_1 + 2a_2 y + \ldots.$$

and

$$\frac{\partial^2 H}{\partial y} = \sum_v v(v-1)y^{v-2} \equiv 1.2a_2 + 2.3a_3 y + \ldots.$$

Substituting the expressions for H, H', H", into the preceding differential equation

$$\frac{\partial^2 H}{\partial y^2} - 2y\frac{\partial H}{\partial y} + (\varepsilon - 1)H = 0$$

and arranging in order of ascending powers of y, gives

$$[1.2a_2 + (\varepsilon - 1)a_0] + [2.3a_3 + (\varepsilon - 1 - 2)a_1]y$$
$$+ [3.4a_4 + (\varepsilon - 1 - 2 \times 2)a_2]y^2 + \ldots = 0$$

For this series to vanish for all values of y, it is necessary that each coefficient of υ vanish for $\nu = 0, 1, 2$, etc.

$$1.2a_2 + (\varepsilon - 1)a_0 = 0 \qquad \nu = 0$$
$$2.3a_3 + (\varepsilon - 1 - 2)a_1 = 0 \qquad \nu = 1$$
$$3.4a_4 + (\varepsilon - 1 - 2.2)a_2 = 0 \qquad \nu = 2$$
$$4.5a_5 + (\varepsilon - 1 - 2.3)a_3 = 0 \qquad \nu = 3$$

A little thought will show that the general rule being followed for the rth coefficient (of y^ν) is

$$(\nu + 1)(\nu + 2)a_{\nu+2} + (\varepsilon - 1 - 2\nu)a_\nu = 0$$

from which we can derive the recursion formula

$$a_{\nu+2} = \frac{-(\varepsilon - 2\nu - 1)}{(\nu + 1)(\nu + 2)}\, a_\nu$$

If we know a_0 and a_1, this expression allows us to calculate all the other coefficients in the power series. a_0 and a_1 are two arbitrary constants that always occur in the solution of an ordinary differential equation of second order.

Left as is, the solution of the S.E. for a harmonic oscillator would not obey the condition that $\psi \to 0$ as $y \to \infty$ since the infinite series would go to ∞ and this would overrun the $e^{-y^2/2}$ asymtotic factor. To fit the boundary condition, the series must be terminated after some finite number of terms. The termination of the series after the vth terms is brought about by the condition

$$\varepsilon = 2\nu + 1$$

This condition will terminate either the series with υ odd or even, not both. Hence, if ν is odd, $a_0 = 0$, and if ν is even, $a_1 = 0$. This equation is a typical eigenvalue condition and reflects the fact that suitable wave functions ψ cannot be found for arbitrary values of the energy.

$$E = \frac{\hbar^2 \varepsilon}{2\alpha^2 m}$$

and with $\alpha^2 = \dfrac{\hbar}{\sqrt{Km}}$ and $\varepsilon = 2\nu + 1$ we get

$$E = \frac{h}{4\pi} \frac{\sqrt{Km}(2\nu + 1)}{m}$$

remembering $K = 4\pi^2 \nu_0^2 m$ simplifies the expresion to

$$E = (\upsilon + \tfrac{1}{2})h\nu_0$$

This is the quantum mechanical expression for the energy levels of a one-dimensional harmonic oscillator. The quantum number υ is a mathematical result of the boundary value condition on the solution of the Schrödinger's equation

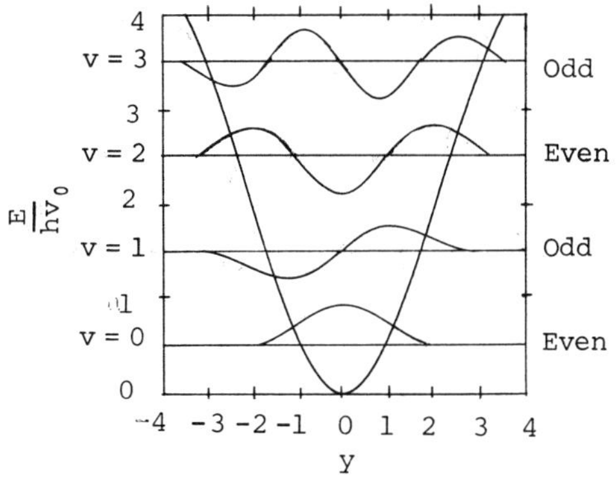

Fig. 24.1

24.2 THE NATURE OF THE HARMONIC OSCILLATOR WAVEFUNCTIONS

The wavefunctions ψ_ν corresponding to the energy levels given by $E = h\nu_0(\nu + \frac{1}{2})$ are the eigenfunctions of the harmonic oscillator.

$$\psi_\nu = N_\nu e^{-y^2/2} H_\nu(y)$$

where N_ν is the normalization factor obtained from the condition

$$\int_{\infty}^{\infty} \psi_\nu^*(x)\,\psi_\nu(x)dx = 1$$

The polynomials $H_\nu(y)$ are known as the Hermite polynomials and can be readily obtained from the definition

$$H_\nu(y) = (-1)^\nu e^{y^2} \frac{\partial^\nu e^{-y^2}}{\partial y^\nu}$$

where the first few are

$$H_0 = 1$$
$$H_1 = 2y$$
$$H_2 = 4y^2 - 2$$
$$H_3 = 8y^3 - 12y$$
$$H_4 = 16y^4 - 48y^2 + 12$$
$$H_5 = 32y^5 - 160y^3 + 120y$$

From these and the asymtotic solution, the normalization condition yields for a wavefunction ψ_ν

$$N_\nu = (\alpha\pi^{\frac{1}{2}}2^\nu \nu!)^{-\frac{1}{2}}$$

$$\int_{-\infty}^{\infty} \psi_1\psi_1^* d\tau = \int_{-\infty}^{\infty} 4e^{-y^2} y^2 dx = \frac{4N_1^2}{\alpha^2} \int_{-\infty}^{\infty} e^{-\frac{x^2}{\alpha^2}} x^2 dx = 1$$

$$2\alpha\pi^{\frac{1}{2}} N_1^2 = 1$$

$$N_1 = \left(\frac{1}{2\alpha\pi^{\frac{1}{2}}} \right)^{\frac{1}{2}}$$

ψ_ν is best interpreted as a probability amplitude the square of which, ψ^2 or $\psi^*\psi$ if the wavefunction has imaginary parts, gives the probability that the particles is within the region x and x + dx. On classical mechanics, an oscillator is most likely to be found at the extreme points of its vibration. The quantum mechanical result distributes the particle over a number of specific regions dependent on the energy of the oscillator. At large enough quantum numbers the particles behavior approaches classical behavior. This is known as the correspondence principle.

24.3 THERMODYNAMICS OF THE HARMONIC OSCILLATOR WAVEFUNCTIONS

The Harmonic oscillator model can be used in the statistical mechanical calculation of the partition function. The thermodynamic functions calculate from Z_v can then be used for the vibration modes of molecules and crystals. The partition function of an individual oscillator is given as

$$Z_v = \sum_v e^{\frac{-(v + \frac{1}{2})h\nu}{kT}} = e^{-\frac{h\nu}{2kT}} \sum_v e^{\frac{-vh\nu}{kT}}$$

where we have used the partition function and substituted for $(v + \frac{1}{2})h\nu$.

From the fact

$$\sum_n^{\infty} x^n = \frac{1}{1 - x} \qquad X << 1$$

$$Z_v = \frac{}{1 - e^{\frac{-h\nu}{kT}}}$$

The total vibrational partition function is the product of terms like this, one for each of the normal modes of vibration.

$$Z_v = \prod_j Z_{v,j}$$

the vibrational energy per mole from

$$\bar{U}_m = RT^2\left(\frac{\partial \ln Z_v}{\partial T}\right)_{V,N}$$

with
$$Z_v = \frac{e^{\frac{-h\nu}{2kT}}}{1 - e^{\frac{-h\nu}{kT}}}$$

becomes

$$\bar{U}_m = 1\,\frac{h\nu}{Z} + \frac{Lhve^{\frac{-h\nu}{kT}}}{1 - e^{\frac{-h\nu}{kT}}}$$

The vibrational heat capacity per mole given as

$$\left(\frac{\partial \bar{U}_m}{\partial T}\right)_V = C_{vm}$$

becomes

$$\left(\frac{\partial \bar{U}_m}{\partial T}\right)_V = C_{vm} = \frac{R\left(\frac{h\nu}{kT}\right)^2}{2(\cosh\frac{h\nu}{kT} - 1)}$$

For the vibrational contribution A = G, since Z_v is not a function of v, P = 0, G = A + PV G = A, and with

$$A = -RT\ln Z_v$$

$$\frac{\bar{G}_m - \bar{U}_{m_0}}{T} = R\ln\left(1 - e^{\frac{-h\nu}{kT}}\right)$$

where U_{m_0} is the zero point energy and is equal to $L\,\frac{hv}{z}$

and with

$$\frac{\overline{U}_m - \overline{U}_{m_0}}{T} = \frac{R\left(\frac{n\nu}{kT}\right) e^{-\frac{h\nu}{kT}}}{\left(1 - e^{-\frac{h\nu}{kT}}\right)}$$

the vibrational contribution to the molar entropy becomes

$$\overline{S} = \frac{\overline{U}_m - \overline{U}_{m_0}}{T} - \frac{\overline{G}_m - \overline{U}_{m_0}}{T}$$

$$\overline{S} = R\left\{\frac{h\nu}{kT}\ \frac{e^{-\frac{h\nu}{kT}}}{1 - e^{-\frac{h\nu}{kT}}} - \ln\left(1 - e^{-\frac{h\nu}{kT}}\right)\right\}$$

24.4 THE RIGID DIATOMIC ROTOR

The motion of two particles of masses m_1 and m_2 formed by a rigid connector of length R, so that the center of mass remains at rest, is equivalent to the motion of a single particle with reduced mass μ at a distance R from the origin. For pure rotational energy there is no potential energy and the Schördinger equation becomes

$$\frac{-\hbar^2}{2\mu} \nabla^2 \psi = E\psi$$

We can change the Laplacian from cartesian to spherical coordinates with r = R a constant, to botain

$$\left(\frac{\partial^2}{\partial\theta^2} + \frac{\cos\theta}{\sin\theta}\frac{\partial}{\partial\theta} + \frac{1}{\sin^2\theta}\frac{\partial^2}{\partial\phi^2}\right)\psi(\theta,\phi) + \beta\psi(\theta,\phi) = 0$$

where $\beta = \frac{2\mu R^2 E}{\hbar^2}$

The variables can be separated to give

$$\psi(\theta,\phi) = \Xi(\theta)\,\Phi(\phi)$$

which when substituted into the Schrödinger equation yields two ordinary differential equations of the form

$$\frac{d^2 \Phi}{d \phi^2} + m_l^2 \Phi = 0$$

$$\frac{d^2 \Xi}{d \theta^2} + \frac{\cos\theta}{\sin\theta} \frac{d\Xi}{d\theta} + \left(\beta - \frac{m_l^2}{\sin^2\theta} \right) \Xi = 0$$

where m_l is an arbitrary separation constant similar to the ones ecountered in the solution of a particle in three-dimensional box. As can be verified by substitution

$$\Phi(\phi) = e^{i m_l \phi}$$

where m_l can take both positive and negative values the restrictions on the wave function and its derivative, single valued, finite, continuous, give

$$\Phi(\phi) = \Phi(\phi + 2\pi)$$

$$e^{i m_l \phi} = e^{i m_l (\phi + 2\pi)}$$

so that $$e^{i m_l 2\pi} = 1$$

for which the only allowable values of m_l are

$$m_l = 0, \pm 1, \pm 2, \pm 3, \ldots.$$

Next, we transform the variable in $\Xi(\theta)$

$$S = \cos\theta$$

$$g(s) = \theta(\cos\theta)$$

since

$$\frac{d\Xi}{d\theta} = -\sin\theta \frac{dg}{ds}$$

$$\frac{d^2\Xi}{d\theta^2} = \sin^2\theta \frac{d^2 g}{ds^2} - \cos\theta \frac{dg}{ds}$$

so that

$$\frac{d^2 \Xi}{d \theta^2} + \frac{\cos\theta}{\sin\theta} \frac{d\Xi}{d\theta} + \left(\beta - \frac{m_l^2}{\sin^2\theta} \right) \Xi = 0$$

becomes

$$(1 - s^2) \frac{d^2 g}{ds^2} - 2s \frac{dg}{ds} + \left(\beta - \frac{m_l^2}{1 - s^2} \right) g = 0$$

The solutions of this well-known differential equation are the associated Legendre polynomials, $P_l^{m_l}(s)$, where the parameter l is related to β by

$$\beta = l(l + 1)$$

As in the case of the Hermite polynomials in the harmonic-oscillator problem, the solutions can be prevented from going to infinity by terminating the series to give a polynomial with a finite number of terms. The condition for such polynomial solutions is

$$l \geq |m_l|$$

The eigenvalue condition on the energy parameter becomes

$$E_l = l(l + 1)\frac{\hbar^2}{2\mu R^2}, \qquad l = 0,1,2,3,\ldots.$$

The eigenfunctions are

$$\psi_{l,m_l}(\theta,\phi) \equiv Y_{l,m_l}(\theta,\phi) = P_l^{|m_l|}(\cos\theta)e^{im_l\phi}$$

for $$m_l = -l,\ -l + 1,\ldots,0,1,\ldots,l$$

For every value of l there are then 2l + 1 different eigenfunctions specified by the values of m_l. The energy levels of the rigid rotor have a degeneracy of 2l + 1.

Surface Spherical Harmonics

l	m_l	$P_l^{m_l}$	In polar coordinates	In Cartesian coordinates	
0	0	1	$f_{00} = 1$	s	$= 1$
1	0	s	$f_{10} = \cos\theta$	p_z	$= z/R$
1	1	$(1-s^2)^{\frac{1}{2}}$	$f_{11} = \begin{cases}\sin\theta\sin\phi \\ \sin\theta\cos\phi\end{cases}$	p_y	$= y/R$
2	0	$\frac{1}{2}(3s^2-1)$	$f_{20} = 3\cos^2\theta - 1$	d_{z^2}	$= (3z^2-R^2)/R^2$
2	1	$3s(1-s^2)^{\frac{1}{2}}$	$f_{21} = \begin{cases}\sin\theta\cos\theta\sin\phi \\ \sin\theta\cos\theta\cos\phi\end{cases}$	d_{yz}	$= yz/R^2$
2	2	$3(1-s^2)$	$f_{22} = \begin{cases}\sin^2\theta\sin^2\phi \\ \sin^2\theta\cos 2\phi\end{cases}$	$d_{x^2-y^2}$	$= (x^2-y^2)/R^2$
3	0	$\frac{1}{2}(5s^2-3s)$	$f_{30} = 5\cos^3\theta - 3\cos\theta$	f_{z^3}	$= (5z^3-3r^2z)/R^3$
3	1	$\frac{3}{2}(1-s)^{\frac{1}{2}}(5s^2-1)$	$f_{31} = \begin{cases}\sin\theta(5\cos^2\theta-1)\sin\phi \\ \sin\theta(5\cos^2\theta-1)\cos\phi\end{cases}$	f_{yz^2}	$= y(5z^2-R^2)R^3$ $= y(5z^2-R^2)/R^3$
3	2	$15(1-s^2)s$	$f_{32} = \begin{cases}\sin^2\theta\cos\theta\sin 2\phi \\ \sin^2\theta\cos\theta\cos 2\phi\end{cases}$	f_{xyz} $f_{z(x^2-y^2)}$	$= xyz/R^3$ $= z(x^2-y^2)/R^3$
3	3	$15(1-s^2)^{3/2}$	$f_{33} = \begin{cases}\sin^3\theta\sin 3\phi \\ \sin^3\theta\cos 3\phi\end{cases}$	f_{y^3} f_{x^3}	$= y(y^2-3x^2)/R^3$ $= x(x^2-3y^2)/R^3$

The function $V_{l,m_l}(\theta,\phi)$ are called surface spherical harmonics.

In the case of diatomic and linear molecules having only one moment of inertia,

$$I = \mu R^2$$

and the eigenvalue condition is written in terms of a rational quantum number J

$$E = J(J+1)\frac{\hbar^2}{2I}$$

24.5 THE THERMODYNAMICS OF THE RIGID ROTOR

The rotational partition function for a linear rigid rotor based on

$$Z = \sum_i g_i e^{-\frac{\varepsilon i}{kT}}$$

is from the eigenvalue condition and the multiplicity of energy levels

$$Z_r = \sum (2J+1)e^{-J(J+1)\frac{\hbar^2}{2IkT}}$$

If the moment of inertia is sufficiently large, the energy levels become so closely spaced that $\Delta\varepsilon$ between adjacent levels is much less than kT even at temperatures of a few degrees Kelvin, and the discrete summation can be repalced by a continuous one

$$Z_r = \int_0^\infty (2J+1)e^{-\frac{J(J+1)\hbar^2}{2IkT}} \, dJ$$

$$Z_r = \frac{2IkT}{\hbar^2}$$

A symmetry number, σ, is introduced into Zr to insure the same states are not counted twice regardless of orientation.

$\sigma = 1$ for heteronuclear molecules and

$\sigma = 2$ for homonuclear ones.

$$Z_r = \frac{2IkT}{\sigma\hbar^2}$$

Consider the calculation of the rotational contribution to molar entropy.

$$\overline{S}_r = RT\,\frac{\partial \ln Z_r}{\partial T} + k\ln Z_r^L = R + R\ln Z_r$$

$$= R + R\ln\frac{2IkT}{\sigma\hbar^2}$$

THE PERIODIC TABLE

KEY: Atomic weight → 112.40; Symbol → **Cd**; Atomic number → 48

METALS / NONMETALS

TRANSITION METALS

PERIODS	IA	IIA	IIIB	IVB	VB	VIB	VIIB	VIII	VIII	VIII	IB	IIB	IIIA	IVA	VA	VIA	VIIA	O
1	1.0079 **H** 1																1.0079 **H** 1	4.00260 **He** 2
2	6.94 **Li** 3	9.01218 **Be** 4											10.81 **B** 5	12.011 **C** 6	14.0067 **N** 7	15.9994 **O** 8	18.9984 **F** 9	20.179 **Ne** 10
3	22.9898 **Na** 11	24.305 **Mg** 12											26.9815 **Al** 13	28.086 **Si** 14	30.9738 **P** 15	32.06 **S** 16	35.453 **Cl** 17	39.948 **Ar** 18
4	39.098 **K** 19	40.08 **Ca** 20	44.9559 **Sc** 21	47.90 **Ti** 22	50.9414 **V** 23	51.996 **Cr** 24	54.9380 **Mn** 25	55.847 **Fe** 26	58.9332 **Co** 27	58.71 **Ni** 28	63.546 **Cu** 29	65.38 **Zn** 30	69.72 **Ga** 31	72.59 **Ge** 32	74.9216 **As** 33	78.96 **Se** 34	79.904 **Br** 35	83.80 **Kr** 36
5	85.4678 **Rb** 37	87.62 **Sr** 38	88.9059 **Y** 39	91.22 **Zr** 40	92.9064 **Nb** 41	95.94 **Mo** 42	98.9062 **Te** 43	101.07 **Ru** 44	102.9055 **Rh** 45	106.4 **Pd** 46	107.868 **Ag** 47	112.40 **Cd** 48	114.82 **In** 49	118.69 **Sn** 50	121.75 **Sb** 51	127.60 **Te** 52	126.9046 **I** 53	131.30 **Xe** 54
6	132.9054 **Cs** 55	137.34 **Ba** 56	57–71 *	178.49 **Hf** 72	180.9479 **Ta** 73	183.85 **W** 74	186.2 **Re** 75	190.2 **Os** 76	192.22 **Ir** 77	195.09 **Pt** 78	196.9665 **Au** 79	200.59 **Hg** 80	204.37 **Tl** 81	207.2 **Pb** 82	208.9804 **Bi** 83	(210) **Po** 84	(210) **At** 85	(222) **Rn** 86
7	(223) **Fr** 87	(226.0254) **Ra** 88	89–103 †	(260) **Ku** 104	(260) **Ha** 105													

*** LANTHANIDE SERIES**	138.9055 **La** 57	140.12 **Ce** 58	140.9077 **Pr** 59	144.24 **Nd** 60	(145) **Pm** 61	150.4 **Sm** 62	151.96 **Eu** 63	157.25 **Gd** 64	158.9254 **Tb** 65	162.50 **Dy** 66	164.9304 **Ho** 67	167.26 **Er** 68	168.9342 **Tm** 69	173.04 **Yb** 70	174.97 **Lu** 71
† ACTINIDE SERIES	(227) **Ac** 89	232.0381 **Th** 90	231.0359 **Pa** 91	238.029 **U** 92	237.0482 **Np** 93	(242) **Pu** 94	(243) **Am** 95	(245) **Cm** 96	(245) **Bk** 97	(248) **Cf** 98	(253) **Es** 99	(254) **Fm** 100	(256) **Md** 101	(253) **No** 102	(257) **Lr** 103